Sneakier Uses for Everyday Things

Sneakier Uses for Everyday Things

How to Turn a Calculator into a
Metal Detector, Carry a Survival Kit
in a Shoestring, Make a Gas Mask
with a Balloon, Turn Dishwashing
Liquid into a Copy Machine, Convert
a Styrofoam Cup into a Speaker,
and Make a Spy Gadget Jacket
with Everyday Things

Cy Tymony

Andrews McMeel
Publishing®

a division of Andrews McMeel Universal

Sneakier Uses for Everyday Things

Andrews McMeel Publishing
a division of Andrews McMeel Universal
1130 Walnut Street, Kansas City, Missouri 64106

www.andrewsmcmeel.com

15 16 17 18 19 TEN 18 17 16 15 14

ISBN: 978-0-7407-5496-8

Library of Congress Control Number: 2005925281

Book design by Holly Ogden

Composed by Kelly & Company, Lee's Summit, Missouri

Attention: Schools and Businesses

Andrews McMeel books are available at quantity discounts with bulk purchase for educational, business, or sales promotional use. For information, please e-mail the Andrews McMeel Publishing Special Sales Department: specialsales@amuniversal.com.

Disclaimer

This book is for the entertainment and edification of its readers. While reasonable care has been exercised with respect to its accuracy, the publisher and the author assume no responsibility for errors or omissions in its content. Nor do we assume liability for any damages resulting from use of the information presented here.

This book contains references to electrical safety that *must* be observed. *Do not use AC power for any projects listed.* Do not place or store magnets near such magnetically sensitive media as videotapes, audiotapes, or computer disks.

Disparities in materials and design methods and the application of components may cause your results to vary from those shown here. The publisher and the author disclaim any liability for injury that may result from the use, proper or improper, of the information contained in this book. We do not guarantee that the information contained herein is complete, safe, or acurate, nor should it be considered a substitute for your good judgment and common sense.

Nothing in this book should be construed or interpreted to infringe on the rights of other persons or to violate criminal statutes. We urge you to obey all laws and respect all rights, including property rights, of others.

Contents

Acknowledgments xi

Introduction xiii

Part I

Sneaky Science Tricks

How to Be Resourceful 3

Sneaky Wire Sources: How to Connect Things 7

Sneaky Work Glove 11

Electroscope 13

Sneaky Hand-Powered Motor 16

Sneaky Hovercraft Toy 18

Sneaky Metal Detector 20

Sneaky Light Sensor 23

Sneaky Light Sensor II 26

Sneaky Flashlight or Laser Beam Communicator 28

Sneaky Speaker 33

Sneaky Speaker II 35

Sneaky Earphone 37

Sneaky Microphone 39

Sneaky Current Tester 41

Antigravity Rollback Toy 43

Sneaky Copier 46

Sneaky Tracer 48

Part II

Sneaky Gadgets

Sneaky Radio-Control Car Projects 53

Sneaky Radio Transmitter and Receiver 58

Sneaky Walkie-Talkie Uses 61

Sneaky Buzzer 66

Sneaky ID Card 70

Part III

Sneaky Survival Techniques

Living on a Shoestring 77

Ink or Swim 80

Sneaky Breathing Device 82

How to Make Invisible Ink 84

Sneaky Invisible Ink II 85

More Hide and Sneak 86

Sneaky Repellent Sprayer 88

Sneaky Security Pen 90

Sneaky Wristwatch 93

Defense or Signal Ring 98

Safety Measure 101

Animal Traps 104

Sneaky Fishing 107

Sneaky Whistle 111

Part IV

Gadget Jacket

Sneaky Interior Pockets 115

Sneaky Sleeve Pockets 120

Sneaky Buttons 124

Sneaky Listener 126

Sneaky Collar 128

Portable Heater 130

Resources 135

Acknowledgments

Special thanks to my agent, Sheree Bykofsky, for believing in the book from the start. I'm also grateful for the assistance provided by Janet Rosen and Megan Buckley at her agency.

I want to thank Katie Anderson, my editor at Andrews McMeel, for her invaluable insights.

I'm grateful to the following people who helped me get the word out about the first "sneaky uses" book:

Gayle Anderson, Sandy Cohen, Ken Hamblin, Deborah Rowe, Ira Flatow, Steve Cochran, Christopher G. Selfridge, Timothy M. Blangger, Cherie Courtade, Charles Bergquist, Mark Frauenfelder, Phillip M. Torrone, M. K. Donaldson, Paul MacGregor, David Chang, Jessica Warren, Steve Metsch, Jenifer N. Johnson, Jerry Davich, Jerry Reno, Austin Michael, Tony Lossano, Katey Schwartz, John Schatzel, Diane Lewis, Bob Kostanczuk, Marty Griffin, Mackenzie Miller, and Rebecca Schuler.

I'm thankful for project evaluation and testing assistance provided by Bill Melzer, Sybil Smith, Isaac English, Jerry Anderson, and Raymond Moore.

And my love goes to my mother, Cloise Shaw, for giving me positive inspiration, a foundation in science, and a love of reading.

Introduction

Ever since the first tool was created, people and societies have been making sneaky uses of everyday things. Whether the adaptation is for novelty purposes or stems from a need for escape and survival, sneaky resourcefulness has produced numerous ingenious innovators. World War II, in particular, inspired many fine examples.

British Royal Air Force pilots were equipped by the Military Intelligence division (MI-9) with various concealed items, such as:

- Shoelaces with magnets in the tips and a wire saw sewn in the fabric
- Compasses and silk maps hidden in buttons and chess pieces
- Boot heels with rubber stamps for document forgery
- Cribbage game boards with crystal radios inside them
- Escape pens that hid a compass, map, currency, and dye to tint clothing

Charles Fraser-Smith—the model for Ian Fleming's character Q (for Quartermaster)—supplied equipment and gadgets for secret agents and prisoners-of-war. Some of his special designs and gadgets included:

- Flashlights with one real battery and a fake with a secret compartment
- Cigarette lighters holding tiny cameras
- Pens containing a paper-thin map, a compass, and a magnetic clip to balance it on a pin
- Buttons containing a tiny compass
- Badges and boot laces containing a Gigli's wire saw (a flexible wire with saw teeth used by surgeons)

Fraser-Smith also developed a used match containing a magnetic needle that could be dropped in water to form a compass, maps printed on handkerchiefs in invisible ink, chess pieces and tobacco pipes with hidden compartments, edible rice-paper notepaper, a cigarette-holder telescope, and fur-lined pilot boots that could be converted into ordinary shoes (to avoid detection) using a knife hidden in the leather, the removed sheepskin legging sections then being converted into a vest).

In Germany's sixteenth-century Colditz Castle, prisoners of war constructed a two-man 19-foot glider with a 33-foot wingspan using cloth from sleeping bags, nails and wood from floorboards, and other materials from their cells.

The inspiration to do this came on a snowy day in December 1943 when prisoner Bill Goldfinch looked out his window over the town and noticed that the snowflakes outside were drifting upward. He thought it might be possible to escape from the old castle in a glider, using the updraft to get airborne.

With the help of a book from the prison library, Goldfinch drew up his specifications. The glider wings would have to have enough lift to carry the glider's pilot and one passenger over the town of Colditz, more than 300 feet below, and across the Mulde River.

In one of the castle's attics, near an adjacent chapel's roof they would use for a runway, the resourceful prisoners created a workshop. With shutters and mud made from attic dust, they constructed a false wall at one end of the attic and went to work, using drills made from nails, saw handles from bed boards, and saw blades from a wind-up record player's spring and the frame around their iron window bars. To cover the glider's wooden frame they used bedsheets, which they painted with hot millet (part of their rations) to stiffen the fabric.

Takeoff was finally scheduled for the spring of 1945. The prisoners planned to assemble the glider and catapult it off the chapel's roof, using a metal bathtub filled with concrete as ballast. The tub, secured to the glider with bedsheet ropes, would fall five stories. The glider would then sail out silently over the town of Colditz, giving its occupants a good head start over the German guards, who would soon discover a bathtub in the yard and two prisoners missing. However, the flight never took place, because the prisoners were rescued by the Americans in 1945. For pictures and more details about the Colditz glider, go to www.sneakyuses.com.

Considering these ingenious contraptions, you can perform amazing feats with the materials you find around you without special knowledge or skills. *Sneaky Uses for Everyday Things* covered such adaptations as how to convert milk into plastic, extract water from air, turn a penny into a radio, and control your TV with a ring. *Sneakier Uses for Everyday Things* goes further and provides more ways to adapt things around you for novel yet practical purposes.

Did you know that you can turn a calculator into a metal detector or store a survival kit in a shoestring? Ever think you could turn a paper cup into a speaker? Adapt liquid detergent into a copy machine? Or make a gas mask out of everyday things? Now you can.

Sneakier Uses for Everyday Things includes science projects, sneaky gadgets, and resourceful survival techniques. No special knowledge or unusual tools are required. Whether your interest is in science or trivia, or you just want to make unique no-cost sneaky gadgetry, you'll undoubtedly look at everyday objects differently from now on.

Get started now—utilize what you've got to get what you want!

Part I

Sneaky Science Tricks

Science is sometimes difficult to understand, but with everyday things, you can make clever animated devices to demonstrate its principles. Many household items you use every day can perform other functions. Using nothing but balloons, paper clips, aluminum foil, paper cups, refrigerator magnets, and other common objects, you can quickly make innovative science projects or demonstration gadgets.

If you are curious about the way static electricity, magnetism, and basic chemistry work, you'll find plenty of project examples here, including an electroscope, a hovercraft, a rollback toy, a sneaky metal detector, an image copier, and various light transmitters and sensors.

Review the sneaky science adaptations in this section, and you'll be ready to create easy-to-make demonstrational projects with items found virtually anywhere.

How to Be Resourceful

The story of the Colditz glider is a great example of the possibilities available to us all if we can adapt everyday things. The key is to think outside the box—to see things as what they can become and not just what you think they're limited to be.

For example, a magazine is an ordinary everyday thing that provides information in printed form, but is that all it's good for? Take a few minutes and think of a periodical's every possible practical application, and then consider the following illustrated examples.

- Remove a staple from a magazine, carefully bend it into a loop shape, dip it in water so a droplet forms on the staple, and you've got a sneaky magnifier.
- Rub a straightened magazine staple ten times in the same direction across a magnet (or a few hundred times against wool cloth or silk material), and it will become magnetized. Then rest it on a floating leaf or piece of wood and one end will point north to create a mini compass.
- If you place a magazine staple across battery terminals, it will heat up enough to ignite tinder material (lint, dried grass, etc.) to start a fire in an emergency. You can also use the magazine as tinder.
- A rolled-up magazine can serve as a funnel to prevent spillage.
- Need a megaphone? Roll a magazine into a cone shape and you can project your voice by speaking into the smaller end.

- A rolled-up magazine pressed against a wall becomes a sneaky sound amplifier.
- With origami folding, a magazine page can become a cup.
- Hide your small flat valuables between pages of a magazine that are glued together.
- Need a defensive weapon? Roll a magazine tightly and jam the end against a person's temple, bridge of the nose, or throat.
- To prevent snow blindness, tear or cut a magazine page into sneaky glasses with slits to look through.
- A magazine can provide insulation when handling hot objects.
- Lost in the cold without sufficient clothing? Tear the pages from a magazine, ball them up, and stuff them in your shirt and pants to provide heat insulation.
- Stand on one end of a rolled-up magazine secured with a rubber band or tape to gain elevation in a pinch.
- Got a flat bicycle tire? Stuff torn magazine pages between the tire and rim to ride home.
- With tape, a magazine page can patch holes in an emergency.
- A very tightly rolled-up magazine can be used as a bottle opener when it is positioned near the neck of a bottle and resting on your thumb (this takes some practice).
- Make a sneaky peashooter barrel with a rolled-up magazine.
- In an extreme emergency, a staple can substitute for a small fuse (*temporarily, not permanently*!).
- If a disk is stuck in a computer CD drive, push a magazine staple into the eject hole to remove it.
- A rolled-up magazine can prop things up such as a window.

magazine

staples

Magnifier

water droplet

staple

Compass

staple

leaf

cup of water

Fire Starter

staple

9 volt

battery

Tinder for Fire

Funnel

Megaphone

Hearing Aid

Cup

(origami)

Safe

Weapon

Snow Glasses

Insulation
(for heat protection)

Insulation
(for cold protection)

Elevator

Flat Tire Filler

Pea Shooter

Bottle Opener

Patch (for opening)

Window Prop

Emergency Fuse

staple

CD-ROM Drive Opener

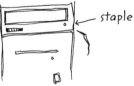

staple

Sneaky Wire Sources:
How to Connect Things

Wire is useful in many sneaky projects. You'll soon learn how it can be used to make a radio transmitter, a speaker, and more.

When wire is required for projects, try whenever possible to use everyday items that you might have otherwise thrown away and help save our natural resources. Common items like potato-chip bags, fast-food wrappers, collector-card packages, and breath-mint labels contain useful aluminum that can be carefully cut to form sneaky wire.

Even a small coffee-creamer container lid, when carefully cut in an up-down-up pattern to utilize its maximum area, can provide a useful connecting wire. See **Figure 1**.

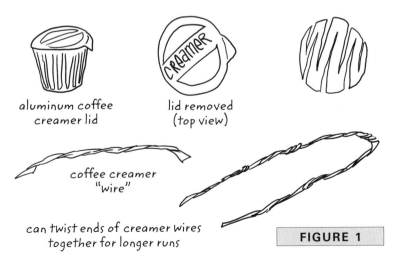

aluminum coffee
creamer lid

lid removed
(top view)

coffee creamer
"wire"

can twist ends of creamer wires
together for longer runs

FIGURE 1

You can test your found wire material for electrical conductivity (to determine if it allows electricity to flow through it) with a battery and either a flashlight bulb or a light-emitting diode (LED). First place the LED leads across the battery terminals to be sure it will light. If not, reverse the direction of the leads. Then place the sneaky wire material in series (end-to-end) with the battery and LED. For example: Press one battery terminal against the "wire" and the other terminal against one of the LED leads. The other LED lead also presses against the wire material. If the LED lights, it's good for using in your sneaky projects. See **Figure 2**.

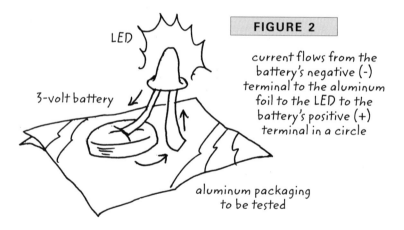

LED

3-volt battery

FIGURE 2

current flows from the battery's negative (-) terminal to the aluminum foil to the LED to the battery's positive (+) terminal in a circle

aluminum packaging to be tested

To insulate your ersatz wires from each other, slip them through discarded straws or wrap paper material around them.

Figure 3 illustrates just a few of the possible items you can use, in case you do not have connecting wire available.

Figure 4 shows how to use a battery and an LED or flashlight bulb to test for electrical conductivity.

Figure 5 shows how to connect wires and LED leads together to insure a good connection.

FIGURE 3

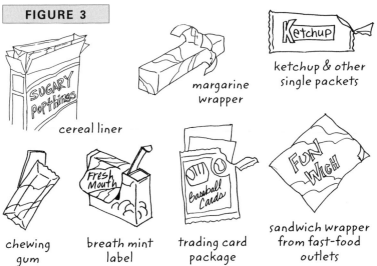

cereal liner

margarine wrapper

ketchup & other single packets

chewing gum

breath mint label

trading card package

sandwich wrapper from fast-food outlets

FIGURE 4

Sample Electrical Circuit: Current flows in circle from negative to device back to positive battery terminal.

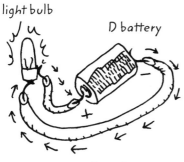

light bulb

D battery

connecting wire

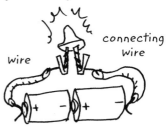

Light Emitting Diode (LED)

wire

connecting wire

2, 1.5 volt batteries in series supply a total of 3 volts

| **FIGURE 5** | *How to Connect Things* |

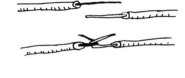

copper wire

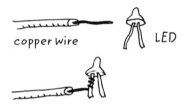
LED

copper wire

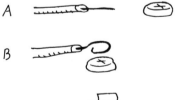

A

B

C

watch battery

Sneaky Work Glove

Work gloves can protect your hands from harm, but they also make it difficult to access small items in your pocket that you may need. You can keep frequently used items with you for quick access with an easy-to-make set of sneaky gloves.

Sneaky gloves use Velcro pads affixed to the back of the small items and on the gloves so you can easily grab items and reattach them on the run.

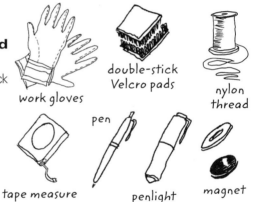

What's Needed
- Work gloves
- Four double-stick Velcro pads
- Nylon thread
- Tape measure
- Pen
- Penlight
- Magnet

work gloves

double-stick Velcro pads

nylon thread

pen

tape measure

penlight

magnet

What to Do
For this project, four common items will be attached to the gloves: a tape measure, a pen, a penlight, and a strong magnet. (The magnet allows you to hold screws, nuts, clips, and other metallic items in place until needed.)

As shown in **Figure 1**, affix half of each Velcro pad to the back of each glove near the wrist area, using its backing tape, or use nylon thread to sew it on. You may want to avoid affixing items near the palm of the glove to avoid scratching surfaces when you lay your hands down.

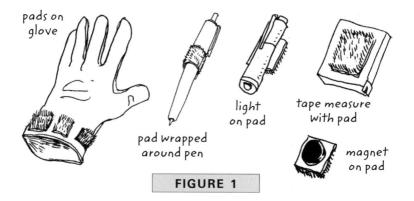

FIGURE 1

Attach the other half of each Velcro pad to one of the four items using its double-stick tape.

After the pads are securely mounted to the items, firmly press the Velcro sides against the pads on the back of the glove, as shown in **Figure 2**.

Now when you work you can take along a light, a pen, a tape measure, and a magnet that will make your next work project easier. If you stick other Velcro pads on a wall or workbench or shelf, you can place gloves and tools there for safekeeping.

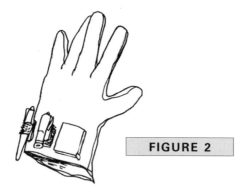

FIGURE 2

Electroscope

You've felt the presence of static electricity when the weather is dry, after receiving a shock when you walk across a carpet and touch someone or a metal object. This static discharge can be powerful enough to damage some electronic items that have sensitive memory chips inside.

You can make a homemade electroscope as a demonstrational device for science projects and for testing for harmful levels of static electricity in your environment.

What's Needed
- Large paper clip
- Two pieces of aluminum foil
- Glass jar with lid
- Quarter (optional)

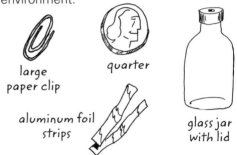

large paper clip

quarter

aluminum foil strips

glass jar with lid

What to Do
All objects, including your body, are a collection of positive and negative electrical particles. Normally there is a neutral state where the positive charges cancel the negative ones. However, in a dry environment, if a charge imbalance, called static electricity, occurs on your body, you can get shocked when you touch a large metal object (or another person). To prevent getting a static electricity shock, touch a doorknob or car door with a coin or key before grabbing it so the spark will emit from the metal instead of your fingertip.

You can make an electroscope easily enough with household items to demonstrate how static electricity charges and discharges objects.

The electroscope consists of two thin pieces of aluminum foil suspended from a metal hook made from a paper clip. When you move the top of the hook near a source of static electricity, some of the electrons in the hook are pushed to the foil and causes them to repel or attract each other.

First, cut two strips of aluminum foil, ⅓ by 1½ inches. Then bend the paper clip into the shape shown in **Figure 1**. Push the hook through the middle of the cardboard bottle cap so the **U** shape protrudes through.

Next, lay the two foil strips one on top of the other and hang them on the end of the hook; see **Figure 2**. Lower the cardboard and the paper clip with foil into the jar so the paper clip is suspended in the center of the jar.

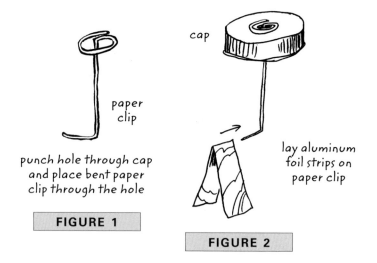

paper clip

punch hole through cap and place bent paper clip through the hole

FIGURE 1

cap

lay aluminum foil strips on paper clip

FIGURE 2

Now hold various metallic and nonmetallic objects in your hand as you walk across the floor (preferably one that's carpeted). Bring the object near the top of the paper clip and observe what happens. You should see the foil strips move apart like little wings. See **Figure 3**.

Then see what happens when the object is moved away from the paper clip. If the strips do not fall back together, gently touch the hook with your finger. *Note:* If you affix a quarter or a large round piece of metal to the top of the paper clip, it can improve the sensitivity of the electroscope.

foil strips fly outward
when static-charged
items are near

FIGURE 3

Sneaky Hand-Powered Motor

As you may know, hot air rises. Rising heat can be made to move objects, and you can demonstrate this fact with a novel "hand-powered" motor. In this demonstrational science project, your hands will actually provide the heat to demonstrate how moving air currents can move an object in a rotary motion.

All it takes is an ordinary piece of paper, scissors, a needle, a cardboard box, and your hands.

What's Needed

- Paper
- Scissors
- Sewing needle
- Small cardboard box

paper

scissors

sewing needle

small box

What to Do

Cut a piece of paper into a 2-inch square. Fold it in half diagonally; then unfold it and fold it in half on the other diagonal, as shown in **Figure 1**. This should create a cross-fold with a center point.

You can use a paper-clip box or similar small box as a mount for the needle. Hold the needle on its side with your fingers and

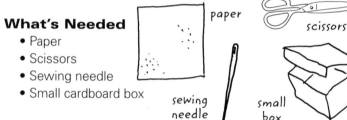

FIGURE 1

fold paper in half   unfold paper   fold over other way

carefully twist it into the top of the box (or use a thimble) until it punctures a hole in the top. Place the piece of paper on top of the needle so its center point allows the paper to turn freely. See **Figure 2**.

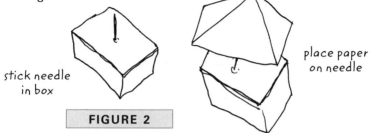

stick needle in box

place paper on needle

FIGURE 2

To make the sneaky "motor" turn, rub your hands together back and forth about twenty times to generate heat and place them near the sides of the paper. After a few seconds, the paper will begin to spin (**Figure 3**).

The paper spins because the heat on your hands causes a temperature increase in the air around the paper. As the heated air rises and cooler air takes its place, the air movement pushes the paper sides, causing it to rotate like a motor.

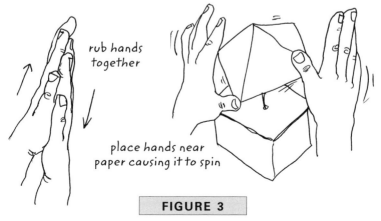

rub hands together

place hands near paper causing it to spin

FIGURE 3

Sneaky Hovercraft Toy

Ever wonder how a hovercraft can ride on air? It floats over surfaces by forcing air underneath its chassis. It also uses large air fans to propel and steer it across the terrain.

Using everyday objects, you can easily demonstrate how compressed air can lift and propel a toy hovercraft vehicle.

What's Needed
- Plastic bottle with top
- Cardboard or CD
- Glue
- Large balloon
- Coffee stirrer
- Tape (plastic or duct)
- Twist-tie

What to Do
The hovercraft will use compressed air stored in a balloon to lift and propel the "vehicle."

Cut off the top of the bottle one-third of the way from the top.

Punch a small hole in the top of the plastic bottle top and tape it to the top of the vehicle and either cut a hole in the center of the cardboard or use a CD and glue the vehicle to the bottle top, as shown in **Figure 1**. The balloon's air will be forced to exit from the bottle top's small hole and provide lift underneath the CD.

Blow up the balloon and place it on the mouth of the bottle. Hold it tight to prevent air from escaping, or wrap a twist-tie around it.

As shown in **Figure 2**, place a coffee stirrer under the lip of the balloon facing the rear of the "hovercraft." Air escaping from the stirrer tube will propel the craft forward.

When you let go of the balloon or unwrap the twist-tie, the air will escape under the hovercraft, causing it to float, and the air from the stirrer will propel it forward, as shown in **Figure 3**.

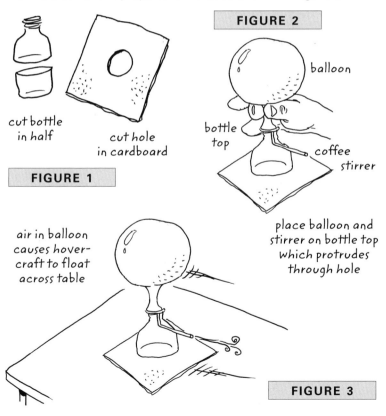

FIGURE 2

balloon

cut bottle
in half

cut hole
in cardboard

bottle
top

coffee
stirrer

FIGURE 1

air in balloon
causes hover-
craft to float
across table

place balloon and
stirrer on bottle top
which protrudes
through hole

FIGURE 3

Sneaky Metal Detector

You've seen metal detectors used in businesses, in public buildings, and in the movies. Believe it or not, you can make a working device using everyday things that can actually detect hidden metal objects.

Many everyday electronic devices—calculators, digital watches, TV remote controls, small video games—emit radio waves that can be detected with an AM radio. This project illustrates the reflective properties of radio waves, using a carefully positioned calculator and an AM radio to detect the radio waves bouncing off a metallic object.

What's Needed

- Small AM radio
- Calculator, solar or battery-powered (or a TV remote control or hand-held video game)
- CD case
- Velcro pads
- Metal object (spoon or tool)
- Aluminum foil
- Tape

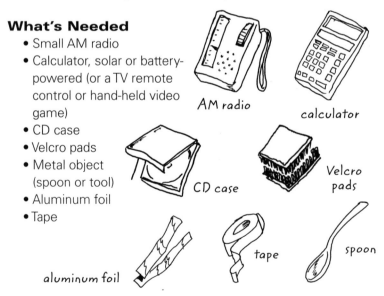

AM radio

calculator

CD case

Velcro pads

aluminum foil

tape

spoon

What to Do

This project requires a radio and a calculator that are small enough to fit on a plastic CD case. Instead of holding the radio and calculator with both hands (and having your body affect the reception), the cover of a CD case allows you to mount them for ease of use. *Note:* You do not need to close the case.

To assemble the sneaky metal detector, affix Velcro pads on the inside cover of the CD case and on the backs of the calculator and the radio. Press the calculator and the radio onto the Velcro pads that are mounted side by side in the CD case. See **Figure 1**. Since CD cases are hinged, you can vary the distance between the radio and calculator.

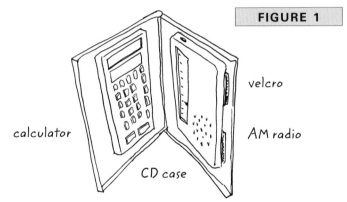

FIGURE 1

velcro

calculator

AM radio

CD case

First, tune the radio near the high end of the AM band, but not directly on a broadcast station. Adjust the volume to the maximum level so you can hear static. Turn on the calculator and position it close to the radio until you hear a loud tone. The tone is the calculator's electronic circuit producing a radio frequency signal. *Note:* Most portable calculators will turn off if no buttons are pressed within a few minutes. If it turns off simply press a key periodically to keep it turned on.

Next, move the radio back until you can barely hear the calculator's tone. While holding the CD case so that the radio and calculator are at about a 90-degree angle, move the case close to a metal object. The tone in the radio should get louder as you get within a few inches of the object. As you move away or to the side, the tone should disappear. Congratulations. You've just made a sneaky metal detector!

With practice, you'll discover the angle position of the CD case that always allows you to hear a tone when it's close to a metallic object, as shown in **Figure 2**.

GOING FURTHER

Test your new sneaky metal detector using other items that emit radio waves, like remote controls and small hand-held electronic games (the kind that are given away at fast-food restaurants).

Also, see if placing aluminum foil between the radio and other items causes the detector to be more (or less) sensitive to the position of metallic objects. Tape the aluminum foil to the back of the CD case to see if it reduces interference from your body and improves performance.

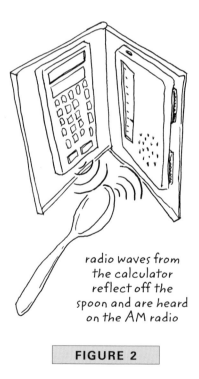

radio waves from
the calculator
reflect off the
spoon and are heard
on the AM radio

FIGURE 2

Sneaky Light Sensor

With a few easy adjustments, a portable cassette recorder (which includes microphone and earphone jacks) can provide you with an audio signal amplifier. You will be able to boost low-level electrical signals and use their increased output for a variety of sneaky applications. All you need is a ⅛-inch plug, the kind you can cannibalize from an old microphone or earphone, and an earphone or speakers.

This project, and the next three, uses a tape recorder to allow you actually to "listen" to light. With a tape recorder's microphone input as an amplifier, you can hear changes in the light around you as sound.

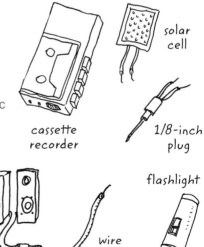

What's Needed

- Cassette recorder
- Solar cell (from an old calculator or other toy; also available at electronic parts stores)
- Cable with ⅛-inch plug
- Earphone or speakers
- Wire
- Flashlight

cassette recorder

solar cell

1/8-inch plug

flashlight

earphone or speakers

wire

What to Do

By connecting a solar cell to the recorder's microphone input
and an earphone or speaker to the microphone jack, you'll be
able to hear changes in light levels.

First, using connecting wire, attach the leads from a solar cell
to a cable with an ⅛-inch plug (using an input plug end from an
old microphone or earphone). Plug the cable into the microphone
input jack of the tape recorder as shown in **Figure 1**.

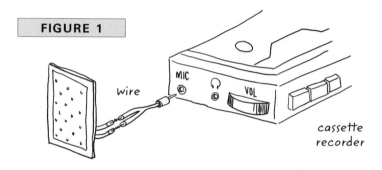

FIGURE 1

MIC

wire

VOL

cassette
recorder

Next, with a blank tape inside the recorder, press the RECORD
button and then the PAUSE button. Placing the recorder in RECORD/
PAUSE mode will activate its amplifier but prevent the tape reels
from spinning to save power.

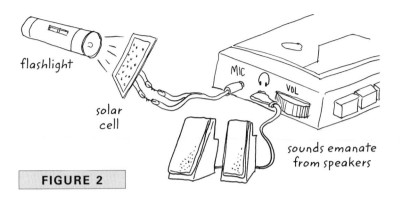

flashlight

solar
cell

MIC

VOL

sounds emanate
from speakers

FIGURE 2

Last, connect an earphone or portable speakers to the
recorder's earphone jack. Move the solar cell around or shine a
flashlight on it, and you will hear various tones. Light from the
solar cell will produce varying voltages and create sounds that
you can hear from the recorder when listening with an earphone
or portable speakers, as shown in **Figure 2**.

Sneaky Light Sensor II

This project will show the versatility of everyday things in science experiments. Light-emitting diodes (LEDs) are found in most electronic devices. They are the little lights that indicate that a device, or function, is enabled.

LEDs, unlike lightbulbs, do not have filaments that heat up to produce light. Instead, an LED is a special diode: an electronic component that conducts electricity through it better in one direction than the other. Unlike regular diodes, LEDs emit light when a low-voltage electrical signal (normally about 2 to 3 volts) passes through it. It operates very quickly, runs cool, and, since it has no filament, theoretically will never burn out.

A little-known fact about LEDs is that they can not only emit light but also sense it. Although this seems similar to a solar cell, an LED, when exposed to a light source, will vary (not produce) an electrical voltage. This will allow you to use an LED for the following sneaky light detection project.

What's Needed

- Cassette recorder
- LED
- Cable with ⅛-inch plug
- Earphone or speakers
- Wire
- Flashlight

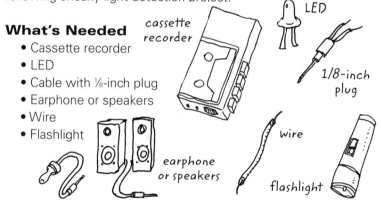

cassette recorder

LED

1/8-inch plug

wire

earphone or speakers

flashlight

What to Do

First, using connecting wire, attach an LED to a cable with an
⅛-inch plug. Plug the cable into the microphone input jack of
a cassette recorder. Then connect an earphone or portable
speaker cable into the earphone jack of the recorder, as shown
in **Figure 1**.

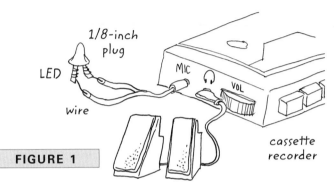

FIGURE 1

Next, place a blank tape in the recorder, press the RECORD and
PAUSE buttons, and set the volume level to maximum. Darken the
room and shine a light on the tip of the LED with the flashlight.
You will hear the static and tone signals from the recorder change
according to the light intensity on the tip of the LED. Test the
sneaky LED sensor near both fluorescent and neon lamps.

GOING FURTHER

Test your sneaky LED light sensor with all the different-color LEDs
you can obtain.

If you place your finger near the tip of the LED, you may find it
will detect the nearness of your body. (Some diodes, even LEDs,
are also sensitive to nearby electrical charges, including those
stored in the human body.)

Sneaky Flashlight or Laser Beam Communicator

Laser beams and LEDs can be used in amazing ways. Most people know that compact discs, DVD players, and bar-code scanners use laser beams to detect and decode electronically encoded digital signals. With special equipment, law enforcement agencies can hear sounds in a room by aiming a laser beam at a window to detect its vibrations.

Now you'll discover how an LED and a battery can be used like a laser beam to transmit and receive sound on a beam of light.

What's Needed

- Cassette tape recorder
- Radio
- Solar cell
- Two AA batteries or one 3-volt watch battery
- Two LEDs, preferably white or yellow
- Flashlight
- Wire
- Tape
- Two cables with ⅛-inch plugs
- Earphone or speakers

cassette recorder

solar cell

radio

LEDs

batteries

flashlight

wire

tape

1/8-inch plugs

earphone or speakers

What to Do

This project requires a sound transmitter and an amplifying receiver device. All necessary parts can be found in discarded toys or purchased at an electronics parts store.

You can build a light transmitter using the parts from an ordinary key ring LED light or by using separate LEDs, batteries, and wire. The sound source can be from a radio or tape player.

This project works best with a white or yellow super-bright LED.

SNEAKY TRANSMITTER

The transmitter consists of a simple battery and LED light circuit connected to a sound source, in this case a radio.

First, connect a wire to both leads of an LED. Then connect the other ends of the wires to the ends of an ⅛-inch plug cable. Next, tape two wires to the ends of the battery and attach them to the LED leads. See **Figure 1**.

Now, plug the ⅛-inch plug cable into the earphone jack of a radio, as shown in **Figure 2**. When tuned to a strong broadcast station, the radio's audio signal will alter (modulate) the light emitting from the LED. This signal can be detected by a light-sensitive device, like a Sneaky Light Sensor receiver device.

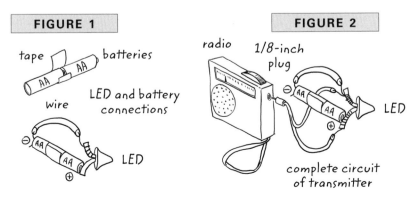

| FIGURE 1 | FIGURE 2 |

tape batteries

LED and battery connections

wire

LED

radio 1/8-inch plug

LED

complete circuit of transmitter

SNEAKY RECEIVER

The light-beam receiver uses the same parts and design as shown in the earlier Sneaky Light Sensor receiver project.

First, connect a solar cell or a white LED to a cable with an ⅛-inch plug and plug it into the microphone jack of the cassette recorder. Then connect the earphone (or speakers) to the earphone jack of the recorder. See **Figures 3 and 4**.

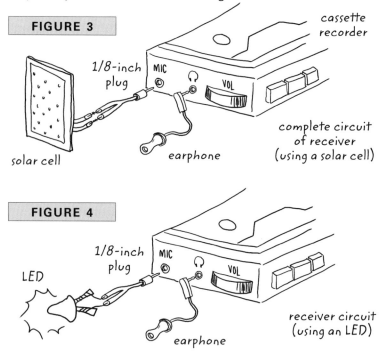

FIGURE 3

cassette recorder

1/8-inch plug

MIC

VOL

solar cell

earphone

complete circuit of receiver (using a solar cell)

FIGURE 4

1/8-inch plug

LED

MIC

VOL

earphone

receiver circuit (using an LED)

Next, place the cassette recorder in RECORD/PAUSE mode to activate the internal amplifier yet prevent the reels from turning to save battery power.

Last, turn up the radio's volume level and position the LED in front of the receiver's solar cell (or LED).

Figure 5 illustrates how to position the receiver's solar cell in front of the transmitter's LED. You can then hear the radio's audio signal from an earphone or speakers plugged into the tape recorder. If necessary, adjust the position and distance of the receiver in relation to the transmitter until a strong signal is detected.

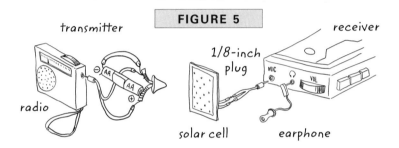

FIGURE 5

transmitter

receiver

1/8-inch plug

radio

solar cell

earphone

GOING FURTHER

This sneaky light-beam sound-transmitter project allows for plenty of sneaky adaptations for experimentation. See the list below and the illustrations in **Figure 6**.

- On the transmitter, connect an additional LED across the first one to increase its range.
- Aim the transmitter through the lens of a magnifying lens or reading glasses to increase the operational distance.
- Set up small mirrors or aluminum foil to see how far you can transmit and receive sound.
- Roll aluminum foil into a tube and test the receiving range.
- Aim a TV remote control at the receiving LED and push its buttons. Can you hear tones from the earphone?

- Perform the same remote control test with the solar cell and compare the distance achieved.
- "See what you can hear" by aiming the receiver at TV and PC screens, fluorescent lights, and other light sources.

FIGURE 6

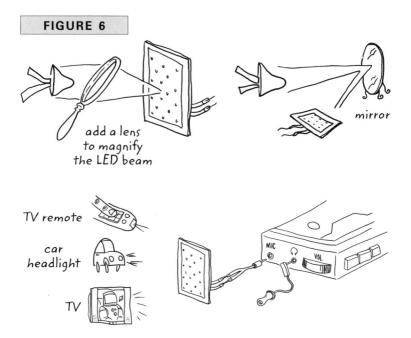

add a lens
to magnify
the LED beam

mirror

TV remote

car
headlight

TV

Sneaky Speaker

If you think you need a big woofer and tweeter or fancy headphones or earphones to hear electronic audio, think again. Speakers convert electrical signals into rapid vibrations to make sound. A typical speaker consists of a coil of wire attached to a paper cone with a magnet mounted close by. When an audio signal travels through the wire, it creates a magnetic field. Since magnets (and magnetic fields) attract and repel each other, the speaker's magnet causes the coil to push in one direction against the paper cone. This rapid motion vibrates the air and creates sound.

This project will show you how to use an ordinary paper or Styrofoam cup, wire, and a magnet to create a sneaky speaker.

What's Needed

- Thick pen or felt-tip marker
- Paper or Styrofoam cup
- Thin wire
- Magnet
- Tape
- Cable with ⅛-inch plug
- Cassette recorder

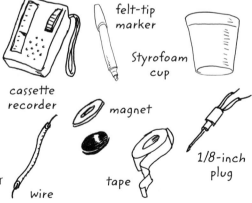

cassette recorder

felt-tip marker

Styrofoam cup

magnet

1/8-inch plug

wire

tape

What to Do

To create a sneaky speaker, wrap ten turns of thin wire around a thick pen or felt-tip marker and use tape to keep it in the shape of a coil. Slide the coil off the pen and tape it to the back of a paper or Styrofoam cup. Tape a small magnet on the back of the coil as shown in **Figure 1**.

FIGURE 1

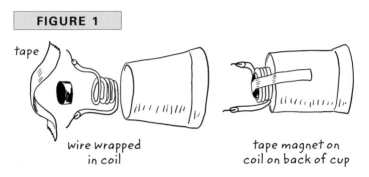

tape

wire wrapped
in coil

tape magnet on
coil on back of cup

Next, connect the coil wires to the cable with an ⅛-inch plug. Insert the plug into the earphone jack on the cassette recorder and turn the volume to maximum, as shown in **Figure 2**. You should be able to hear sounds from the cup. If not, reposition the magnet on the wire coil. For a louder volume, use a larger magnet.

FIGURE 2

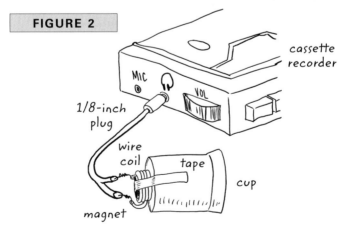

cassette
recorder

MIC

VOL

1/8-inch
plug

wire
coil

tape

cup

magnet

Sneaky Speaker II

In the first Sneaky Speaker project, you learned that a speaker consists of a coil of wire attached to a paper cone with a magnet mounted close by. Believe it or not, another common everyday item can become a sneaky speaker, further demonstrating the amazing versatility of wire and magnets.

What's Needed

- Electric motor (from a toy car)
- Paper or Styrofoam cup
- Cable with ⅛-inch plug
- Cassette recorder
- Wire

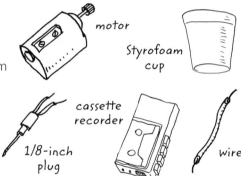

motor

Styrofoam cup

cassette recorder

1/8-inch plug

wire

What to Do

Electric motors convert electrical power into motion using a coil of wire and magnets. Normally you supply battery power to get a mechanical spinning motion from it. You can obtain a different effect by feeding the audio output of a radio or tape player into it and using a paper cup to "amplify" the vibrations for you to hear.

You can remove a motor from a discarded toy car for this project demonstration. As shown in **Figure 1**, connect the two wires of an electric motor from a toy car to an ⅛-inch plug cable. Insert the plug into the earphone jack on the radio, and turn the volume to its maximum level.

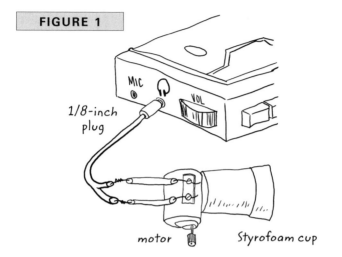

FIGURE 1

MIC

1/8-inch plug

VOL

motor   Styrofoam cup

The audio signal from the radio creates a magnetic field in the motor's coil windings that repel against the motor's internal magnets, causing it to vibrate. With the motor lying on a table, rest a paper or Styrofoam cup on it so it's positioned at a near-45-degree angle. The cup will vibrate when resting on the motor and allow you to hear sounds from the radio. If not, reposition the cup on the motor until the sound is audible.

Sneaky Earphone

Your outer ear is always collecting air vibrations for your inner ear to process into sound. Using the same items in the Sneaky Speaker project, without the cup, you can turn your outer ear into a "speaker" and experience more of the sneaky versatility of a simple piece of wire.

What's Needed

- Magnet
- Thin wire
- Tape
- Cable with ⅛-inch plug
- Pen or felt-tip marker
- Cassette recorder

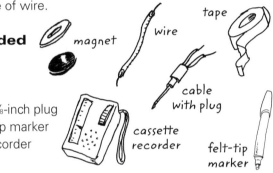

magnet

wire

tape

cable with plug

cassette recorder

felt-tip marker

What to Do

First, wrap about ten turns of thin wire around a thick pen or felt-tip marker and use tape to keep it in the shape of a coil. Slide the coil off the pen and tape and connect the coil wires to the ⅛-inch plug cable wires.

Next, insert the plug into the earphone jack on the radio, as shown in **Figure 1**, and turn the volume to maximum. Press the wire coil to the back of the cup of your ear and press the magnet near the coil.

The magnet will cause the coil to vibrate your outer ear, and you'll actually hear the audio from the radio. *Note:* Even with the radio at high volume, there is no harm to your ear in performing this project because of the very weak signal from the coil of wire.

FIGURE 1

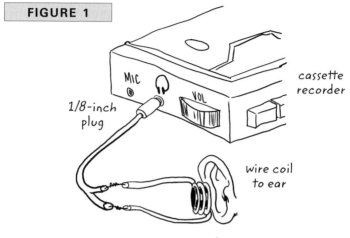

magnet

Sneaky Microphone

The Sneaky Speaker and Sneaky Earphone projects have demonstrated how speakers convert an electrical signal into sound by vibrating the air. This project shows how to reverse this effect to make a Sneaky Microphone.

What's Needed

- Cassette recorder
- Paper or Styrofoam cup
- Magnet
- Thin wire
- Tape
- Cable with ⅛-inch plug

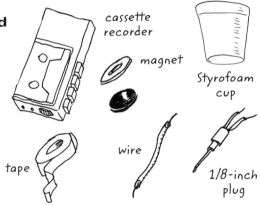

cassette recorder

magnet

Styrofoam cup

wire

tape

1/8-inch plug

What to Do

Just as an electrical signal in a coil attached near a magnet can vibrate a cup and produce sound, so can you reverse the effect and create a Sneaky Microphone using the same parts and setup as the Sneaky Speaker.

By speaking loudly into the cup, you will vibrate the coil. When the coil vibrates near the magnet, it produces an electrical signal that corresponds to your voice. The electrical signal is detected and amplified by the tape recorder.

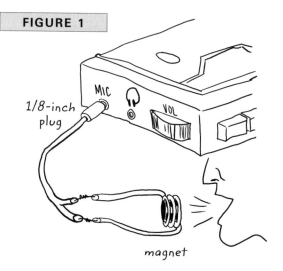

FIGURE 1

MIC

1/8-inch plug

VOL

magnet

Plug the ⅛-inch plug cable into the microphone input jack on the cassette recorder. With a blank tape in the recorder, you can record your voice (press just the RECORD button) and listen to it by playing back the tape or press the RECORD and PAUSE buttons and listen with an earphone or speakers connected to the earphone jack.

Sneaky Current Tester

Here's a sneaky fact: Small motors and even fruits can generate electrical power. You can make a current tester with everyday things to see this phenomenon.

What's Needed

- Wire
- Compass
- Clear tape
- Two paper clips
- Small electric motor
- Lemon, battery, or solar cell (optional)

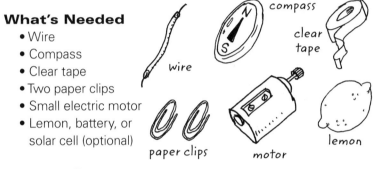

compass

clear tape

wire

paper clips

motor

lemon

What to Do

First wrap eight turns of connecting wire around the compass and secure it with tape, as shown in **Figure 1**. Remove a motor from a small motorized toy and disconnect its two wires from the toy's main board. Connect the two motor wires to the wires around the compass.

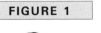
FIGURE 1

wrap wire around compass

Then bend a paper clip in the crank shape shown in **Figure 2** and attach it to the motor's gear. The paper clip allows you to turn the gear like a crank. By rotating the crank you can produce an electrical current strong enough to cause the compass needle to move.

Besides the electric motor, the Sneaky Current Tester can also detect other sources of electrical power—obtained from a lemon or battery or solar cell—as shown in **Figures 2 and 3**.

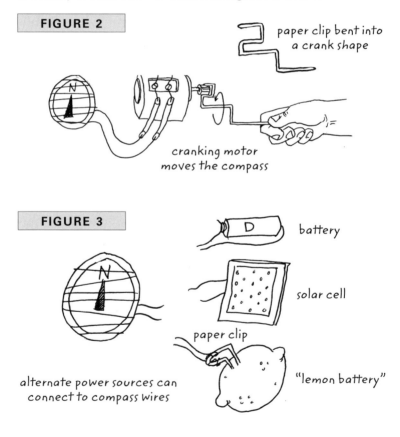

FIGURE 2

paper clip bent into a crank shape

cranking motor moves the compass

FIGURE 3

battery

solar cell

paper clip

alternate power sources can connect to compass wires

"lemon battery"

Antigravity Rollback Toy

If you've wondered how hybrid cars can boast such impressive mile-per-gallon ratings, this project will show you how. You will demonstrate the principle that allows hybrid cars to store and release energy by using a simple-to-build rollback toy.

What's Needed

- Small container with plastic lid
- Extra plastic lid
- Thick rubber band 3–4 inches long
- Two paper clips
- Bolts or large lug nuts
- Scissors

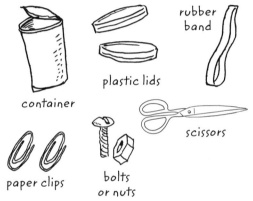

rubber band

plastic lids

container

scissors

paper clips

bolts or nuts

What to Do

First, obtain a small cardboard container and remove the bottom. Cut slits through the center of the can's plastic lid and the extra lid. Take the lids off the container. See **Figure 1**. Thread a rubber band through the bottom of the container and pull it through the lidless top of the container.

Tape the washers or lug nuts together and connect them to the middle of one section of the rubber band (do not tape the strands of the rubber band together). See **Figure 2**.

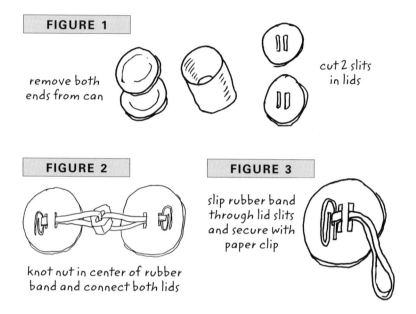

FIGURE 1

remove both
ends from can

cut 2 slits
in lids

FIGURE 2

knot nut in center of rubber
band and connect both lids

FIGURE 3

slip rubber band
through lid slits
and secure with
paper clip

Put the end of the rubber band through the container lid. Use a paper clip to secure the band so it does not slip inside the container (put the paper clip through the end loop of the rubber band that remains outside the hole). See **Figure 3**. Put the lid on the container, making sure the rubber band is still sticking out the other end.

Carefully pull on the rubber band until it comes through the hole. Secure the band with the second paper clip. Be sure to situate the weight so it is in the center of the container and does not touch the sides. Put both lids on the ends of the container. See **Figure 4**. Your rollback toy is ready to go!

Roll the toy and watch as it stops and returns to you. See **Figure 5**. This sight is more amazing when you roll it downhill and it stops and returns to you uphill, seemingly defying gravity.

FIGURE 4

bend and push one
lid through can and
put lids on both ends

FIGURE 5

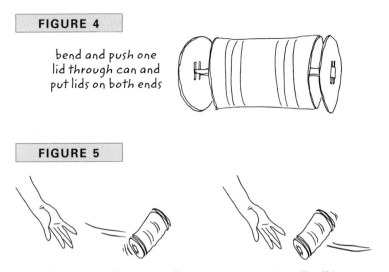

rolling away from you will store energy and it will roll back

The weight holds one end of the rubber band stationary while the free side twists around. The farther the toy rolls, the more potential energy is stored. Release and watch the toy roll back toward you, demonstrating its conversion into kinetic energy.

Hybrid vehicles use this principle to store energy in a flywheel to power an electric motor. The motor is used when you take off from a standing stop to save engine fuel.

Sneaky Copier

Have you ever found yourself needing to make a copy of a drawing and no copy machine is around? Using household items you can make a sneaky copy machine of your own

What's Needed

- Teaspoon
- Vanilla extract
- Liquid dish detergent
- Small bowl
- White paper
- Ink pens of various colors
- Paintbrush (optional)
- Comic strip or newspaper picture (optional)

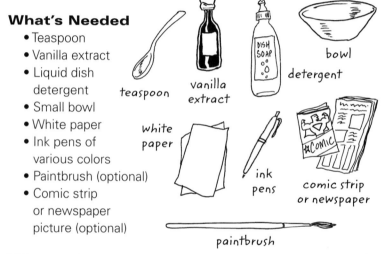

teaspoon

vanilla extract

bowl

detergent

white paper

ink pens

comic strip or newspaper

paintbrush

What to Do

First, draw a picture on a sheet of paper, using a black pen. Go over the lines to thicken them. (Thicker lines allow you to make more copies from the original.)

Next, mix equal parts of vanilla extract and liquid dish detergent together in a small bowl; one teaspoon of each should be enough. Using your finger or a small paintbrush, completely cover the drawing with a thin layer of your Sneaky Copier solution, as shown in **Figure 1**.

FIGURE 1

mix extract and dishwashing
liquid in bowl

rub "copier" liquid
mix on original

Now place a clean sheet of white paper on top of the picture. Rub the back of the paper firmly with the bowl of the teaspoon until the picture begins to show through the paper. Peel the paper off the picture to see your copier creation; see **Figure 2**. You should be able to make multiple copies this way if your original drawing has thick dark lines.

Last, test the Sneaky Copier technique with other ink colors. Try to make a copy of a newspaper page or a comic strip. Keep in mind that the images on the copy will be reversed.

FIGURE 2

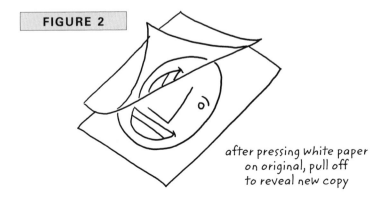

after pressing white paper
on original, pull off
to reveal new copy

Sneaky Tracer

Some items cannot be copied using a chemical transfer technique (shown in the preceding Sneaky Copier project), because their images are printed on coated paper or, in the case of text, the image will be reversed.

Another way to make a duplicate of an original image is to use a Sneaky Tracer.

What's Needed

- Four cardboard strips
- Hole punch or nail
- Paper clips
- Paper-clip box
- Tape
- Two pencils

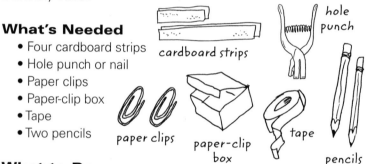

cardboard strips

hole punch

paper clips

paper-clip box

tape

pencils

What to Do

First, cut two pieces of cardboard, each measuring 2 by 8 inches. Then cut another two pieces, each 2 by 4 inches.

Arrange the cardboard pieces in the pattern shown in **Figure 1** and then punch holes in the corners of the shape. Bend paper clips into **C** shapes and push them through the holes to secure the pieces, yet still allow them to move freely.

Next, punch pencil-sized holes in the cardboard at points A and B, just large enough so a pencil can fit through snugly, and insert a pencil in each (see **Figure 2**). Now place the copier device so one end rests on the top of the paper-clip box and secure it with tape. The box acts as an elevated mounting

FIGURE 1

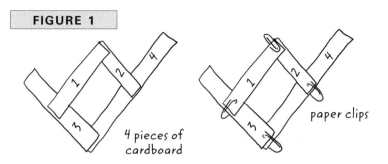

4 pieces of cardboard

paper clips

platform to keep the pencils balanced and stable yet free to move about.

Last, select an original drawing that you want to trace and set it under the pencil in hole A. Place a blank sheet of paper under hole B. Use pencil A to trace the drawing and you'll see another picture being created by pencil B. If necessary, secure the pencils to the cardboard and the paper to the table with tape. See **Figure 3**.

Now you can easily trace complex drawings and make copies for your needs. Experiment with the lengths of cardboard, and you'll see that you can easily enlarge or reduce the size of the drawings made.

FIGURE 2

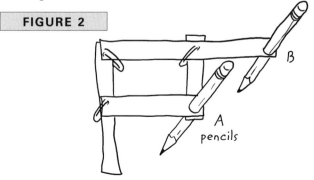

B

A
pencils

FIGURE 3

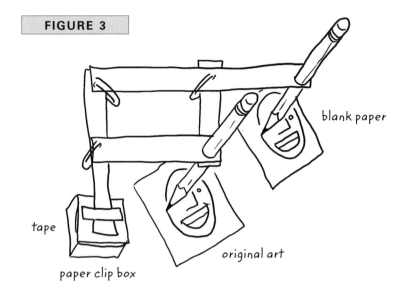

blank paper

tape

paper clip box

original art

Part II

Sneaky Gadgets

If you're curious about the sneaky adaptation possibilities of household gadgets, you have the right book. People frequently throw away damaged gadgets and toys without realizing they can serve unintended purposes.

It's hard to believe, but such common household items as broken tape recorders, LED flashlights, key-chain voice memo recorders, radios, walkie-talkies, and toy car motors can be used to create a novel ID card, a radio transmitter made with only a piece of wire, inventive radio-controlled car applications, sneaky walkie-talkie uses, and more.

All these projects are tested safe, and you can make them in no time. If you enjoy the idea of high-tech resourcefulness, the following chapters will undoubtedly provide plenty of resourceful ideas.

Sneaky Radio-Control Car Projects

Radio-controlled cars have many sneaky adaptation possibilities that can increase their usefulness. This project uses the inexpensive single-function type of radio-controlled toy car; this model will travel forward continuously, once its ON/OFF switch is placed in the ON position, until you actuate the remote control button, causing it to back up and turn. When you release the control, the vehicle goes forward in a straight line again.

The instructions and illustrations that follow will show you how to modify the transmitter to a more compact size, to use it as an alarm trigger. You'll also see how to modify the receiver to activate other devices, such as lights and buzzers.

What's Needed

- Radio-controlled car
- Three 3-volt watch batteries (or fewer, depending on transmitter)
- LEDs
- Buzzer
- Tape
- Wire
- Rubber band
- Playing card
- Strong thin thread

radio-controlled car

watch batteries

LEDs

buzzer

tape

wire

rubber band

playing card

thread

What to Do

How a radio-controlled car works. Pressing the transmitter button closes an electrical switch, which turns on the transmitter. This sends electromagnetic waves through the air that are detected by the radio receiver in the vehicle. The receiver detects the radio signal from the transmitter and reverses the electrical polarity (direction of current flow) of the power applied to the motor. This causes it to run in the reverse direction.

Adapting the car's transmitter. The first sneaky adaptation to the transmitter is to make it as small as possible, for concealment inside other objects or clothing.

Since the transmitter is always in the OFF mode until its activator button is pressed, it can operate using tiny long-life watch batteries.

If the transmitter uses one AA or AAA battery, it can be replaced by one small watch battery with the same voltage output. *Note:* Each AA or AAA battery supplies 1½ volts of power.

If the transmitter operates on two AA or AAA batteries, you can substitute either two 1½-volt watch batteries or a single 3-volt watch battery. If a 9-volt battery was in use, you will need to use three 3-volt watch batteries. When stacking batteries, place the positive side of one battery against the negative side of the other.

Figure 1 shows how to replace regular AA or 9-volt batteries in the transmitter with 3-volt watch batteries.

If you connect two wires across the transmitter's activator button, you can have another sensor or switch activate the transmitter to alert you of an entry breech or that your valuables are being removed. Place a piece of tape over the transmitter button so that when the device is activated it will be on. See Figure 2.

Adaptating the car's receiver. You can also modify the car's radio receiver, which is on a circuit board in the car's body, for use as an alarm trigger. See Figure 3.

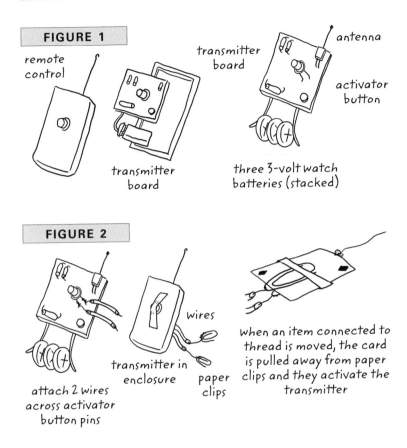

FIGURE 1

remote
control

transmitter
board

antenna

transmitter
board

activator
button

three 3-volt watch
batteries (stacked)

FIGURE 2

wires

transmitter in
enclosure

paper
clips

attach 2 wires
across activator
button pins

when an item connected to
thread is moved, the card
is pulled away from paper
clips and they activate the
transmitter

Unlike the transmitter, the receiver must stay ON to be able
to operate, and this produces a small constant drain on the
batteries. **Figure 4** illustrates how to modify the receiver for use
with watch batteries using the same technique described for the
radio transmitter.

If desired, the toy car motor can be used in an application of
your own design. (*Sneaky Uses for Everyday Things* illustrated

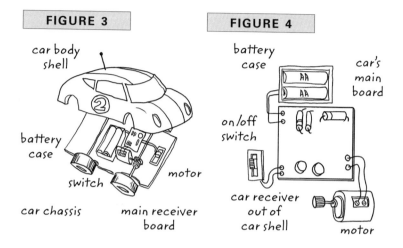

FIGURE 3

car body shell
battery case
switch
car chassis
main receiver board
motor

FIGURE 4

battery case
car's main board
on/off switch
car receiver out of car shell
motor

a Door Opener using a toy car.) The car motor is attached to the receiver with two connecting wires. If you physically remove the motor from the car body (either by unclipping or unscrewing it), you can use the receiver for more project applications. It's easy to connect the receiver's motor wires to other devices to activate them remotely.

Figure 5 shows how the wires in the receiver that previously connected to the motor can be connected to other devices, like an LED or a buzzer for remote control.

FIGURE 5

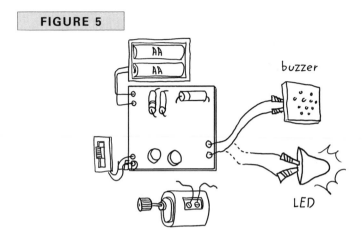

Sneaky Radio Transmitter and Receiver

Have you ever seen movie characters rig a radio transmitter to save the day? What does it take to make a radio transmitter—transistors? integrated circuits?

Believe it or not, there's an easy, sneaky way to transmit audio privately from a radio, TV, cassette, or MP-3 player to a nearby person's radio or tape recorder, with just a loop of wire!

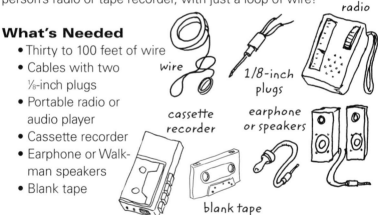

radio

wire

1/8-inch plugs

cassette recorder

earphone or speakers

blank tape

What's Needed
- Thirty to 100 feet of wire
- Cables with two ⅛-inch plugs
- Portable radio or audio player
- Cassette recorder
- Earphone or Walkman speakers
- Blank tape

What to Do
A wire loop creates a magnetic field when an audio signal is connected to it. Another wire loop, located within or near the original wire loop, can detect the magnetic field from the other wire loop.

When the smaller wire loop, acting as a Sneaky Receiver, is connected to the microphone input of a tape recorder, you can hear the original sound signal.

SNEAKY RADIO TRANSMITTER

The transmitter consists of a single loop of wire. Use a small loop of wire, about 4 feet in length, to start and experiment with longer wire lengths later.

First, connect the two ends of the wire loop to the ends of the ⅛-inch-plug cable.

Next, plug the ⅛-inch plug cable into the earphone jack of a radio or other audio device, as shown in **Figure 1**. When tuned to a strong broadcast station, the radio's audio signal will alter (modulate) a magnetic field in the wire loop. This signal can be detected by another wire loop connected to an amplifier.

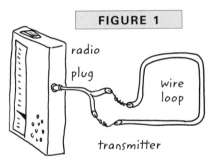

FIGURE 1

radio

plug

wire loop

transmitter

SNEAKY RECEIVER

The light-beam receiver uses a design similar to the earlier Sneaky Light Sensor project, but, instead of a solar cell, a wire loop is used as an input device.

First, connect the ends of a 2-foot length of wire to both ends of a cable with an ⅛-inch plug. Plug the cable into the microphone input jack of the tape recorder and connect an earphone or portable speakers to the recorder's earphone jack, as shown in **Figure 2**.

Next, with a blank tape inside the recorder, press the RECORD button and then the PAUSE button. Placing the recorder in RECORD/PAUSE mode will activate its amplifier but prevent the tape reels from spinning to save power.

Last, turn up the radio's volume level and move the wire loop close to the radio transmitter's larger loop, and you will hear the radio's broadcast station. See **Figure 3**.

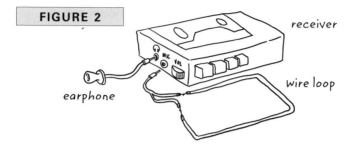

FIGURE 2

receiver

earphone

wire loop

GOING FURTHER

If necessary, adjust the position and distance of the receiver in relation to the transmitter until a strong signal is detected. Experiment with a larger loop and see how far away you can get from the transmitter and still receive a signal.

Use a 50-foot length of wire for the transmitter and lay it around the perimeter of a room. Test to see if you can walk around the room and listen to the radio using the portable tape recorder.

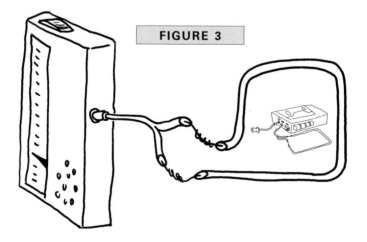

FIGURE 3

Sneaky Walkie-Talkie Uses

A pair of compact walkie-talkies lends itself to a variety of sneaky applications. Some models are so small, they are now mounted in toy wristwatches. This project illustrates methods to use walkie-talkies as an intercom, as an alarm sensor, and as a sneaky listening device.

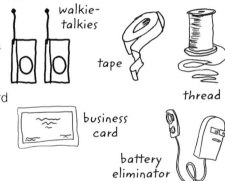

What's Needed
- Pair of walkie-talkies
- Tape
- Nylon thread
- Coated business card or bookmark
- Nine-volt battery eliminator (optional)

What to Do

First, test the walkie-talkies with fresh batteries and note their maximum reliable operating distance. Follow the directions in the categories below for your desired application.

SNEAKY INTERCOM

This is the easiest project application to set up. It takes the place of an intercom system when wires cannot be placed between locations. For example, you can place one walkie-talkie in one room of the house and the other in the basement, the garage, a bedroom, or at the front door (mounted securely with screws or glue underneath a protective awning).

Simply mount the first walkie-talkie in a remote area (possibly under a cover to protect it from the elements) and listen with the second unit as shown in **Figure 1**.

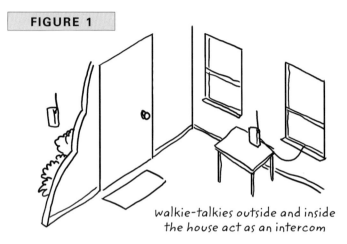

FIGURE 1

walkie-talkies outside and inside the house act as an intercom

For constant monitoring applications, (e.g., as a baby monitor), apply a piece of strong tape across the TALK button so the unit is always in transmit mode. This application will eventually drain the battery, so you may want to use walkie-talkies that use 9-volt batteries. Then you can attach an AC battery eliminator to the battery clip (available at electronic parts stores), so it can always be in the ON or STANDBY mode without requiring frequent battery replacement. See **Figure 2**.

SNEAKY LISTENER

Use the Sneaky Listener application, similar in operation to the Sneaky Intercom, when you want to monitor a remote location

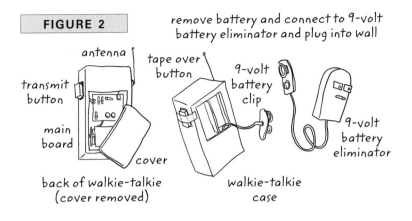

FIGURE 2

remove battery and connect to 9-volt
battery eliminator and plug into wall

antenna

tape over
button

transmit
button

9-volt
battery
clip

main
board

9-volt
battery
eliminator

cover

back of walkie-talkie
(cover removed)

walkie-talkie
case

secretly. Simply place one walkie-talkie out of sight, within or under an object. Place the power button in the ON position and put tape across the TALK button to keep it transmitting.

As shown in **Figure 3**, you can monitor the audio in the area with the other walkie-talkie (and retrieve the remote one later). Or, as shown in the Gadget Jacket project (Part IV), a walkie-talkie can be placed in a jacket (that's left in a room) to monitor nearby sounds from afar.

SNEAKY ALARM TRIGGER

A walkie-talkie set provides an inexpensive quick-to-set-up option for a sneaky wireless alarm system. One walkie-talkie set up with tape across its TALK button, to keep it in the transmit mode, can broadcast a warning signal to the remote unit. The trick is to place an insulator strip, from a coated business card or a bookmark, between one battery terminal and its clip. Connect a thin strong nylon thread or wire to the other end of the insulator strip and wrap it around the item you want to protect.

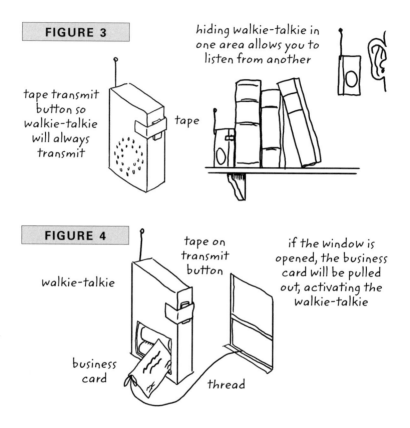

FIGURE 3

hiding walkie-talkie in one area allows you to listen from another

tape transmit button so walkie-talkie will always transmit

tape

FIGURE 4

tape on transmit button

walkie-talkie

if the window is opened, the business card will be pulled out, activating the walkie-talkie

business card

thread

First, open the first walkie-talkie and remove one side of the 9-volt battery clip so that it rests on top of the battery terminal. Then poke a small hole in the insulator strip and tie the thread through it into a knot. Attach the other end to a window handle, door knob, or other object that you want to keep from being moved.

Next, place the insulator strip between the battery terminal and the battery clip. The battery clip should still be attached to the other battery terminal and should keep pressure on the insulator. If the insulator is pulled away from the battery, the battery clip should rest on the top of the battery terminal and turn on the walkie-talkie. If not, wrap a small rubber band around the battery and the battery clip, to apply more pressure. See Figure 4.

Turn on the walkie-talkie's power button and tape the transmit or TALK button so it stays on. Place it out of sight from the window or object that it will be connected to with the thread.

When someone opens the window, the thread will pull the insulator away from the battery clip, turning on the walkie-talkie, which will transmit to the other walkie-talkie.

Note: If the walkie-talkie includes a signal button or a Morse code signaler, it can be taped in the ON position too.

Sneaky Buzzer

A buzzer is a fairly common item. Actually, it's a simple electromagnet with a clever modification that demonstrates a very elegantly designed circuit. You can make your own Sneaky Buzzer with found items like paper clips, wire, and a battery.

What's Needed

- C- or D-size battery
- Twenty-five feet of wire
- Magnet
- Fifteen paper clips
- Cardboard
- Tape

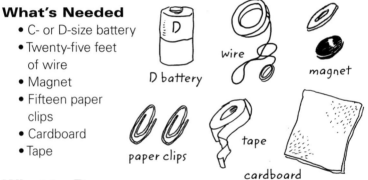

D battery

wire

magnet

paper clips

tape

cardboard

What to Do

When electricity flows through a wire, it creates a magnetic field around it. Winding wire in a coil and placing a piece of metal inside the coil increases the strength of the field and creates an electromagnet.

A buzzer consists of an electromagnetic circuit with a piece of metal close by that is pulled onto another piece of metal and creates a tapping sound. The metal is actually a part of the wire coil circuit; when it is pulled toward the second piece, it turns off the electromagnet.

The metal then returns to its original position because of spring tension. Once in place again, the electromagnetic field is

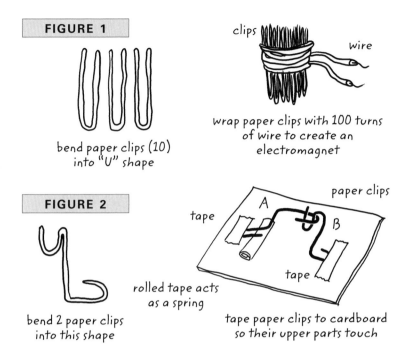

FIGURE 1

bend paper clips (10)
into "U" shape

clips
wire

wrap paper clips with 100 turns
of wire to create an
electromagnet

FIGURE 2

bend 2 paper clips
into this shape

tape
A
B
paper clips

rolled tape acts
as a spring

tape

tape paper clips to cardboard
so their upper parts touch

activated and the cycle restarts. The rapid tapping of the metal
on the second metal piece is what causes the buzzing sound.

To make an electromagnet, wind a coil of wire around **U**-
shaped paper clips, as shown in **Figure 1**.

Then bend two paper clips into the shape shown in **Figure 2**
and mount them on the cardboard with tape so they touch.
Notice that paper clip A rests on a roll of tape, which acts as
a spring to keep it in contact with paper clip B.

Next, tape one coil wire to paper clip A and the other to one
battery terminal with tape. Connect the other battery terminal
to paper clip B. See **Figure 3**. In this arrangement, the battery's

FIGURE 3

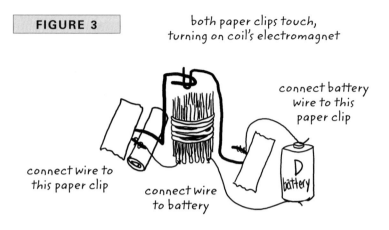

both paper clips touch,
turning on coil's electromagnet

connect battery
wire to this
paper clip

connect wire to
this paper clip

connect wire
to battery

D
battery

power flows from one of its terminals to the wire coil, through both paper clips, and back to the second battery terminal. In this circuit it does not matter which direction, positive or negative, the battery terminal is connected to the other parts.

Once the battery is connected, the wire coil electromagnet will attract paper clip A toward it and move away from paper clip B. This disconnects the electromagnetic circuit, and the electromagnet turns off. The rolled-up tape "spring" pushes the paper clip up toward the second paper clip and makes a tapping sound. When the two paper clips are in contact again, the coil turns on and the cycle begins again. See **Figure 4**.

FIGURE 4

when the wire coil has power, it attracts paper clip and disconnects battery

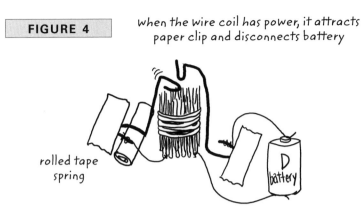

rolled tape spring

once disconnected, the paper clip is pushed back up by the "spring" tension; this connects both paper clips, turning on the electromagnet

Sneaky ID Card

If you want the safety and convenience of a "card-swiping" security system in your home (like the ones used at many businesses), you can use a few household objects to achieve your dream.

With a few components, you can create a sneaky personal ID card, allowing you to identify your friends through walls and doors.

What's Needed

- Cardboard
- Wire
- Two paper clips
- Two C-size batteries
- LED or buzzer
- Glue
- Tape
- Two business cards

wire

paper clips

cardboard

LED

C batteries

glue

tape

What to Do

The Sneaky Card ("Don't sneak around without it") is designed to activate a complementary signal circuit that will properly identify you as a "friendly" carrier.

The Sneaky Access Card is made of two ordinary business cards, or a card of your own design, with two bent paper clips inside. Two small loops of the paper clips will protrude through the back of the card so they can make contact with the signal circuit.

To make the card, simply straighten the paper clips into a **U** shape and poke them through one card about two inches from each other. Bend the paper clips back onto the card so they rest flat and touch each other. See **Figure 1**. Then glue the second card over the first so it conceals the paper clips connections, as shown in **Figure 2**.

FIGURE 1

front   back

punch 2 paper clips through card and bend back

FIGURE 2

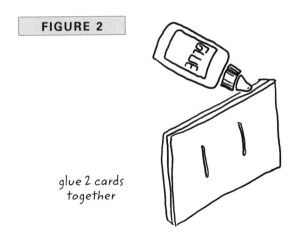

glue 2 cards
together

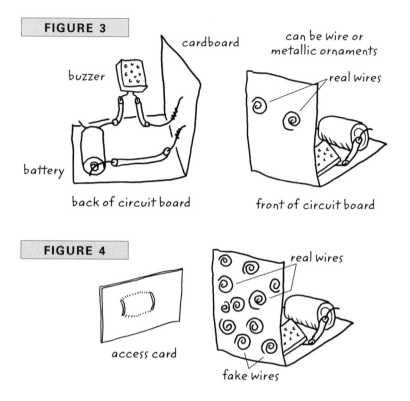

FIGURE 3

cardboard

can be wire or
metallic ornaments

buzzer

real wires

battery

back of circuit board

front of circuit board

FIGURE 4

real wires

access card

fake wires

The signal circuit consists of a battery and a buzzer, with two coils of wire acting as a switch. When a piece of metal contacts both wire coils, the buzzer will turn on. The items are easily mounted and assembled on a piece of cardboard and are secured with tape. The front of the signal circuit stands vertically with the small wire coils protruding through the front. See **Figure 3**. Additional wire coils can be placed on the front of the signal circuit board for subterfuge. See **Figure 4**.

When you place the Sneaky Access Card on the surface of

the signal circuit in the correct position, it completes the circuit and activates the buzzer. See **Figure 5**. For your own applications you can extend the wires going to the signal-circuit front panel so it's on the other side of a door or other partition.

access card

pressing access
card against wires
turns on buzzer

Part III

Sneaky Survival Techniques

We prefer to believe that misfortunes won't occur, but why take chances? Why not prepare for the worst when you're traveling? With the proper knowledge, you can improvise sneaky gear that may save you and others from harm.

Most people would object to carrying a full array of emergency equipment, but you can have the most fundamental items available in a package as small as a shoelace. Who can argue against that?

After reviewing this section, you'll be equipped, among other things, to assemble a shoestring survivor kit; make a sneaky breathing device; hide security devices in a ring, pen, or watch; construct animal traps and fishing lures with items in your pocket; and make a whistle with a blade of grass.

Living on a Shoestring

When you take a hiking or camping trip, it's wise to bring emergency provisions. Despite such preparations, sometimes you might leave your main camp and get separated from your bag or backpack. If you get lost, what will you do? Why not carry the most essential items on your person, so even if you get separated from your group, you'll always have basic equipment with which to save your life.

You can make a sneaky survival kit that fits into the most unlikely places, including a jacket patch, a pen, a watch, even your shoestrings! Even if you never have to remove your Sneaky Survival Kit while hiking or camping, you will never be without it.

What's Needed

The list below contains the most essential items you should have in case of emergencies. The quantity of items depends on the amount of room that's available in your desired storage compartment.

- Two aspirins
- Penlight bulb
- Watch battery
- Wire (to connect bulb and battery for flashlight)
- Three-inch square of aluminum foil (as signal mirror)
- Dental floss (for fishing)
- Multivitamins
- Duct tape

aspirin

penlight bulb

watch battery

dental floss

tape

- Wooden match heads
- Paper clip
  (to hold match heads)
- Small Baggie
- Antibacterial tablet
  (to treat water in Baggie)
- Survival information sheet
  (semaphore signals, Morse
  code, etc.)
- Pencil lead (for notes on
  survival sheet)
- Mints (for emergency
  sugar source)
- Needle
- Magnet (to make compass
  from a needle or paper clip)

What to Do

As shown in **Figure 1**, you can remove the inner parts from a
large pen or an inexpensive watch and carefully position all these
items inside. Roll the aluminum foil, survival instruction sheets,
and plastic Baggie tightly into a very small size.

FIGURE 1

Or you can very carefully sew all of the items between two jacket patches and sew the patch onto your jacket. It's also possible to arrange all the survival items carefully between two athletic shoestrings and then sew them together with the items in between as shown in **Figure 2**. This saves the space in your pen or watch for larger essentials: currency, coins—and balloons and rubber bands for Sneaky Water Wings!

FIGURE 2

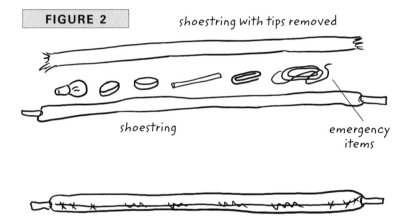

shoestring with tips removed

shoestring

emergency items

shoestrings sewn together with emergency items inside

Ink or Swim

Who among us has not feared for our small companions, children and pets alike, when they're near a body of water? With a few everyday items, you can prevent a possible drowning. Just bring this easy-to-make lifesaving gadget with you when traveling over or near water with your little ones.

What's Needed

- Large wide-barreled pen
- Four large balloons (two for each arm)
- Four large rubber bands

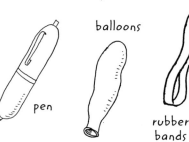

balloons

pen

rubber bands

What to Do

You can store emergency sneaky water wings available for nonswimmers inside a large hollow pen barrel. When tightly wrapped, the balloons and rubber bands will easily fit inside a pen, as shown in **Figure 1**. Just slip the pen in a pocket or purse before a trip for added peace of mind.

Figure 2 shows how to affix the sneaky water wings to a child's arms in case of an emergency swim. If the rubber bands are too large for the arms or legs of the child (or pet), just wrap them around multiple times for a snug fit.

FIGURE 1

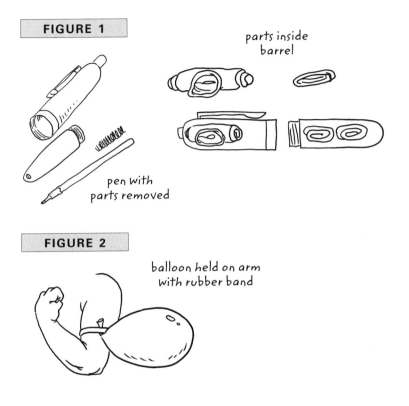

parts inside
barrel

pen with
parts removed

FIGURE 2

balloon held on arm
with rubber band

Sneaky Breathing Device

Did you know that smoke inhalation causes more fire-related deaths than flames or heat? The smoke uses up oxygen required for breathing and replaces it with toxic gases that can quickly make you drowsy and disoriented.

If you must enter an area clouded with thick smoke, you need to continue breathing clean air as long as possible. Most people can hold their breath for just under a minute. Here is an inexpensive and easy way to have clean air for another five to ten minutes, long enough for you to move to safety or for rescuers to arrive.

What's Needed

- Compact air pump
- Large balloons
- Rubber bands (optional)
- Flashlight (optional)

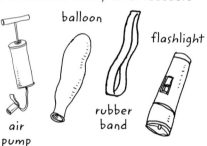

What to Do

First, always keep a flashlight with fresh alkaline batteries handy in your house or office in case of emergencies. Let everyone know its location for easy access. Attach several balloons and a compact air pump to the flashlight with a large tightly wrapped rubber band. This will allow you to grab all three items quickly, even in the dark. In case the lights go out, the flashlight will allow you to signal for help at a window.

If you must remain in a room that is slowly filling with smoke or if you absolutely must enter a smoke-filled hall, pump up the balloons with clean air and wrap the ends tight around your finger or wrap rubber bands around them to prevent air from escaping. See **Figure 1**. Take a deep breath, hold it as long as possible, and exhale. Inhale clean air from the balloons as needed.

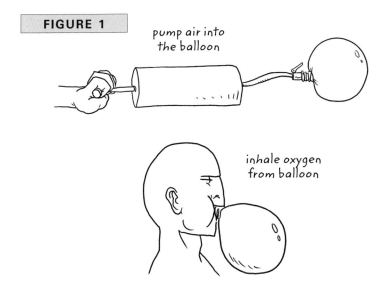

FIGURE 1

pump air into the balloon

inhale oxygen from balloon

How to Make Invisible Ink

If anything is a prime example of a Sneaky Use project, it's using everyday things to make invisible ink. (Sneaky fact: Casinos now use cards marked with symbols that are only visible when viewed with a special lens.) You can use a large variety of liquids to write secret messages. In fact, some prisoners of war used their own saliva and sweat to make invisible ink.

What's Needed

- Milk *or* lemon juice *or* equal parts baking soda and water
- Small bowl
- Cotton swab or toothpick
- Paper

bowl

cotton swabs

lemon juice

paper

What to Do

Use a cotton swab or toothpick to write a message on white paper, using the milk or lemon juice or baking soda solution as invisible ink. The writing will disappear when the "ink" dries.

To view the message, hold the paper up to a heat source, such as a lightbulb. The baking soda will cause the writing in the paper to turn brown. Lemon or lime juice contain carbon and, when heated, darken to make the message visible.

You can also reveal the message by painting over the baking soda solution on the paper with purple grape juice. The message will be bluish in color.

Sneaky Invisible Ink II

Here's another sneaky method to write and view invisible messages that stay invisible (unless you know the trick), this time using common laundry detergent and water.

What's Needed

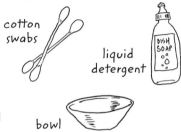

cotton swabs

liquid detergent

bowl

- Cotton swab or towel
- Small bowl
- Liquid detergent
- Water
- Black light
- Piece of white cardboard

What to Do

In a small bowl, mix a teaspoon of liquid laundry detergent with one cup of water and use a small towel or cotton swab to write a message on a white piece of cardboard. See **Figure 1**. The message will not be visible at this point.

To view the secret message, darken the room and shine a black light—invisible ultraviolet light—on the board. The previously invisible message will become visible, as shown in **Figure 2**.

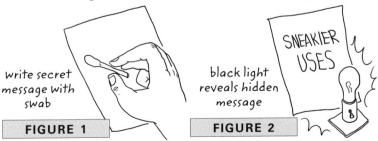

write secret message with swab

FIGURE 1

black light reveals hidden message

SNEAKIER USES

FIGURE 2

More Hide and Sneak

When you think of sneaky you usually think of something that is secret or hidden from you. Actually, the most common sneaky-use application is hiding your valuable belongings from others.

The following figures show how to keep your things to yourself, even if they are in plain sight. Most likely a thief or nosy houseguest will briefly examine the item and then ignore it as a possible safe. See **Figure 1** for examples.

- Hollowed-out candle
- Figurine
- Tissue container
- Trash container base
- Video or audio cassette shell
- Pen
- Watch case
- Inner pocket
- Shoestring
- DVD case
- Between magazine pages
- Inside a candy box
- Ironing board padding
- Bag within a bag

FIGURE 1

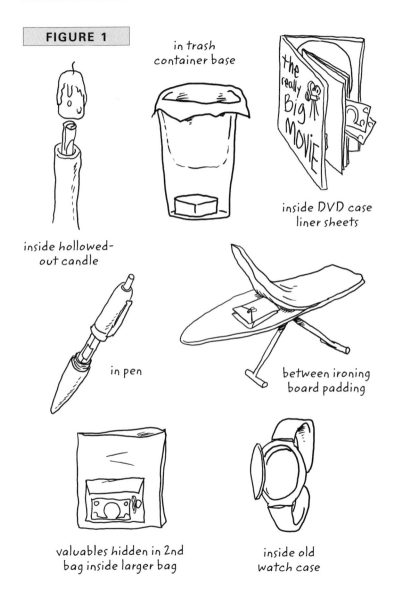

in trash
container base

inside DVD case
liner sheets

inside hollowed-
out candle

in pen

between ironing
board padding

valuables hidden in 2nd
bag inside larger bag

inside old
watch case

Sneaky Repellent Sprayer

If you're wary of a predator (animal or human) approaching you, you can carry your own concealed defense device made with household items. The Sneaky Sprayer provides you with a safe deterrent that wards off potential attackers.

You can carry the sprayer in your pocket or bag. Or you can attach it to the Gadget Jacket's Sleeve Device (see Part IV).

What's Needed

- Small squeeze bottle
- Baby powder
- Ground cayenne pepper

squeeze bottle

cayenne pepper

baby powder

What to Do

When you perceive a potentially dangerous situation, it's always best to leave the area. But in some situations you may need a deterrent to keep others away and give you precious moments to escape. (It's also important not to cause any harm to anyone in the process.)

The Sneaky Repellent Sprayer uses a combination of two safe ingredients: baby powder and pepper.

First, find a small squeeze bottle (the kind that trial-size shampoos and hand sanitizers come in) with a flip-up top. Empty the container, rinse it thoroughly, and allow it to dry.

Next, unscrew the top and fill the bottle one-quarter full of baby powder, another quarter with pepper, and a third quarter with baby powder. See **Figure 1**. Replace the top, and you're ready.

FIGURE 1

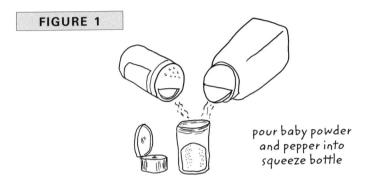

pour baby powder and pepper into squeeze bottle

You may want to test the Sneaky Repellent Sprayer outside to see how hard you should squeeze and how far the repellent mixture will travel, as shown in **Figure 2**.

Let's hope you never need this deterrent, but you can be prepared without spending a dime, using these household items.

FIGURE 2

sprayer in action

Sneaky Security Pen

When you are away from home, you may need a quick way to set up sneaky security devices to alert you to unauthorized entry into your room via door or windows. You may be in a hotel or at a friend's house and want to have sneaky noisemakers go off to alert you when you're sleeping or in another room.

Using everyday items stored inside a pen, you can sleep soundly, knowing your windows and doors cannot be tampered with without your knowledge.

What's Needed

- Large wide-barreled pen
- Two large rubber bands
- Two small chimes
- Fifteen feet of dental floss
- Duct tape
- Piece of thin bubble wrap, 2 by 4 inches

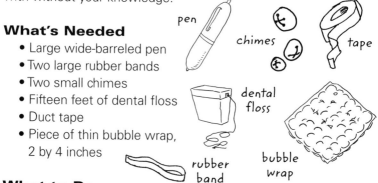

What to Do

First, unscrew the pen and remove its inner parts. Then carefully insert the rubber bands, chimes, and floss into the pen. You can also put a small sheet of bubble wrap (the thin type used to protect small objects) around the other items, tape it, and insert it into the pen. **Figure 1** shows the essential parts.

To set your sneaky booby trap, attach one end of the floss around a door or window handle (or both) and wrap the other end around a chair or desk leg. Then tape one chime in the

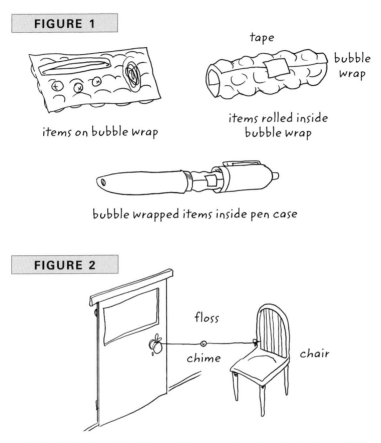

FIGURE 1

tape

bubble wrap

items on bubble wrap

items rolled inside bubble wrap

bubble wrapped items inside pen case

FIGURE 2

floss

chime

chair

middle of the length of dental floss, so that it will alert you if the door or window is opened. See **Figure 2**.

Next place rubber bands, linked if necessary, around the doorknob and a nearby object, like a window handle, to impede entry and tape a chime to it. It won't prevent someone from entering the room but it will cause the chime to ring repeatedly. See **Figure 3**.

FIGURE 3

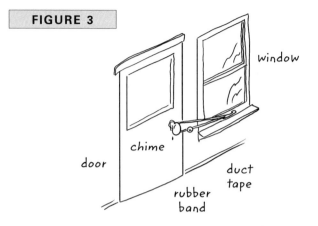

window

chime

door

duct
tape

rubber
band

Last, set the small piece of bubble wrap near the door (or under a mat if there is one) to alert you by popping if someone enters and walks on it. See **Figure 4**.

FIGURE 4

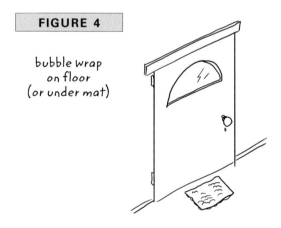

bubble wrap
on floor
(or under mat)

Sneaky Wristwatch

Some situations require a rapid response and quick access to the proper tools to be effective. The most convenient place to store a sneaky gadget is in a wristwatch. Because it's always at hand, you don't have to waste time fumbling for it in a pocket or bag. Also, wristwatches are so common they don't attract unnecessary attention, and devices hidden inside can be used secretly.

This project explains how to assemble a Sneaky Wristwatch to store covert devices for a variety of applications.

What's Needed

- Wristwatch
- Voice recorder
- Siren
- Radio-control transmitter
- Retractable key ring
- Mini-light
- Watch batteries

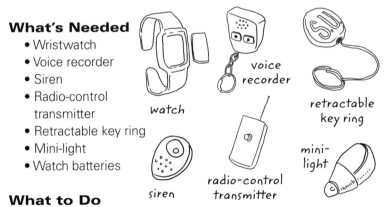

watch

voice recorder

retractable key ring

siren

radio-control transmitter

mini-light

What to Do

First, obtain the largest toy wristwatch you can find that looks real. Depending on your needs, you can remove the watch parts and install a miniature voice recorder, a simple radio transmitter (from a radio-controlled car), an electronic mini-siren, a retractable key ring, or any other small gadget that will fit in the watch case. See **Figure 1**.

Next, locate the smallest version available of the item you will mount in the case. Most drugstores, department stores, and toy stores sell miniature inexpensive versions of the foregoing devices for just a few dollars.

A compact device like a voice recorder usually uses tiny watch batteries for power. But a device like a radio-control transmitter may require two AA batteries. If the device uses AA or AAA batteries, you can substitute tiny watch batteries with the same voltage output. *Note:* Each AA or AAA battery supplies 1½ volts of power.

For instance, if the device uses one AA battery, use one 1½-volt watch battery in its place. If two AA or AAA batteries are needed, you can substitute either two 1½-volt watch batteries or a single 3-volt watch battery.

Figure 2 shows how to remove a tiny voice recorder from its original case and exchange its large 6-volt battery for two smaller 3-volt watch batteries, stacked together. (When stacking batteries, place the positive side of one battery against the negative side of the other.)

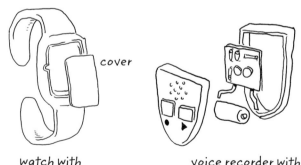

| **FIGURE 1** | **FIGURE 2** |

cover

watch with
cover removed

voice recorder with
cover removed

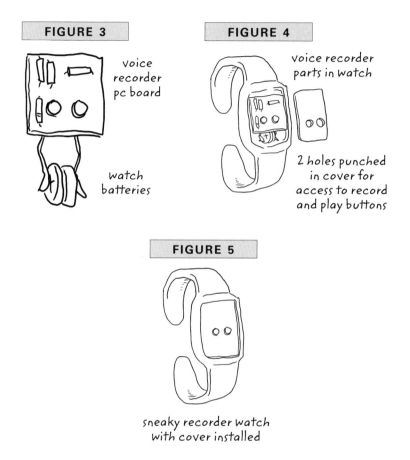

FIGURE 3

voice
recorder
pc board

watch
batteries

FIGURE 4

voice recorder
parts in watch

2 holes punched
in cover for
access to record
and play buttons

FIGURE 5

sneaky recorder watch
with cover installed

First remove the inner parts of the watch (circuit board and batteries) and put them aside. Then carefully position the desired device inside the case. In the example shown in **Figures 3 and 4**, a small voice-recorder device is placed inside of the empty watch case and small holes are punched in the watch face for access to the recorder's PLAY and RECORD buttons. See **Figure 5**.

FIGURE 6	FIGURE 7	FIGURE 8

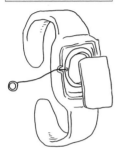

hole punched in
watch case; key ring
placed in watch case

connect 2 extra
ultra-bright
LEDs to mini-
lights battery

parts inside
watch case

GOING FURTHER

The sneaky technique of emptying the watch case and punching
appropriate holes in its front face allows for innumerable items
to be secreted inside, including:

- A retractable key cord (**Figure** 6)
- A mini-keychain light, with additional LEDs across its watch
  battery for almost blinding brightness (**Figures 7, 8, and 9**)
- A radio-control transmitter (from a toy car) or mini-siren
  (shown in **Figures 10, 11, and 12**)

FIGURE 9

pushing button
emits blinding light

FIGURE 10

FIGURE 11

siren p.c. board
in watch case

FIGURE 12

activating
siren

Defense or Signal Ring

Here's a compact gadget you can make that will serve three practical purposes. It's a mini-flashlight or an emergency safety beacon, and it can act as a sneaky self-defense device.

What's Needed

- Large toy ring with a tin band and bubble top
- Ultra-bright LED key-ring light
- Two extra ultra-bright LEDs
- Three-volt watch battery
- Wire
- Glue
- Paper
- Transparent tape

toy ring

light

LED

watch battery

wire

glue

paper

tape

What to Do

Select an adjustable toy ring that has a clear case mount so the LED light can shine through. Or obtain a clear bubble top from another ring, toy, or ornament and glue it to your ring later.

Remove the battery and ultra-bright LED from the key-ring light. See **Figure 1**. Connect the LED from the toy and the two other LEDs to one piece of wire as shown in **Figure 2**.

Next, position the other LED leads to one side of the watch battery, as shown in **Figure 3**. Ensure that all the LEDs light when you touch the other end of the wire to the other side of the battery. If not, reverse the leads and test them again.

Place insulated material, like tape or vinyl or paper, on the

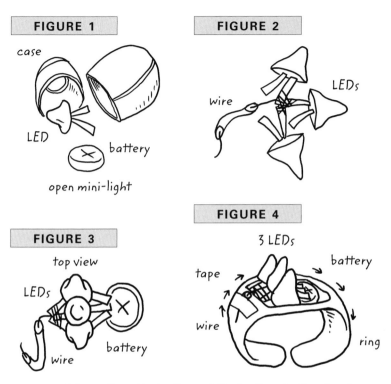

FIGURE 1

case

LED

battery

open mini-light

FIGURE 2

wire

LEDs

FIGURE 3

top view

LEDs

wire

battery

FIGURE 4

3 LEDs

tape

battery

wire

ring

current flows from the bottom of the battery to the ring, to the wire, then to the LEDs and back to the top of the battery

surface and on one side of the ring. Then set the battery and LEDs on top of the ring's surface. Lead the other end of the wire to the outside of the ring (**Figure 4**). In this manner the insulating material prevents the bottom of the battery from contacting the ring's metal surface and inadvertently turning on the LEDs.

Last, when all parts are positioned properly, glue the toy bubble top back on and press the wire end against the tin band.

FIGURE 5

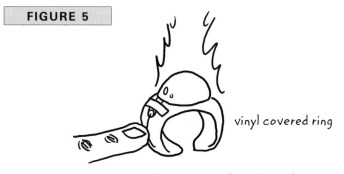

vinyl covered ring

pressing wire on side activates ultra-bright light

This will connect the wire to the battery and turn on the LEDs, causing them to emit blinding light. See **Figure 5**.

Your sneaky ring can now serve as a convenient flashlight, as an emergency safety blinker (when walking at night in unlit areas), and as a self-defense device to temporarily blind and disorient an attacker.

Safety Measure

Traveling brings a sense of adventure, discovery—and, some-times, danger. While staying away from home you can gain a sense of security easily by setting up a sneaky safety alarm, using just one or two everyday things.

What's Needed

- Retractable mini tape measure or key cord
- Tape
- Small chime (optional)

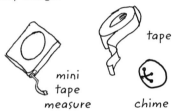

mini tape measure

tape

chime

What to Do

Since they both extend and recoil with a noticeable sound, a retractable compact tape measure or key cord provides a versatile sneaky tool for sensing when a door or window has been breeched. It can also alert you when an item has been inappropriately moved from the surface of a desk.

Simply extend the tape measure and carefully place its end tab under a window handle or a doorknob. See **Figures 1 and 2.** If it won't stay in position, use tape to secure it properly.

Want to know if someone moves one of your valuables? Simply place the tape measure's end underneath it and let it extend down the other side of a table, as shown in **Figure 3.** When the item is lifted, the tape measure will retract with a loud snap, alerting you and catching the person off guard.

For an extended measure of safety, you can set up a tape measure to turn on a noisemaker, like a portable battery-powered

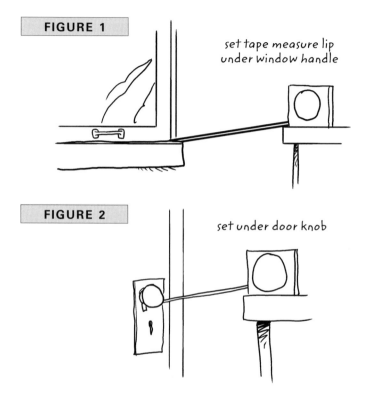

FIGURE 1

set tape measure lip
under window handle

FIGURE 2

set under door knob

radio. As shown in **Figure 4**, place a piece of tape between a battery and its connector in the radio. Simply attach the tape to one end of the tape measure and secure the radio and tape measure so they will not move.

When the window is opened, the end of the tape measure slides back to its case, pulling the tape away from the battery and causing the radio to turn on and alert you.

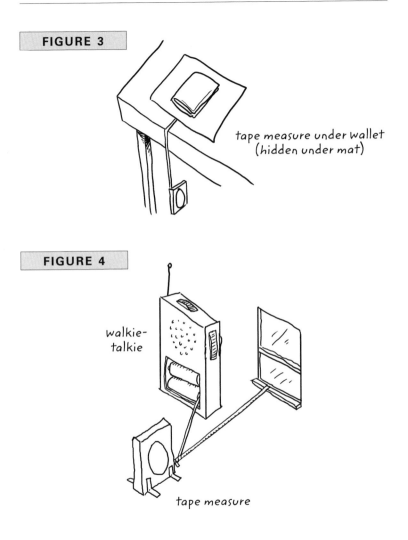

FIGURE 3

tape measure under wallet
(hidden under mat)

FIGURE 4

walkie-
talkie

tape measure

Animal Traps

If you're ever trapped in the wilderness, capturing small animals for food can prove to be energy-wasting and very difficult. Using items you may already have in your pocket, along with rope, twigs, string, twine, and wire, you'll be able to set small sneaky traps to catch emergency food.

By setting several traps, you'll have the ability to catch much more potential food than by chasing animals with a weapon (which you probably don't have anyway!).

What's Needed

- Box
- Sticks
- Wire
- Strong threads from clothing or string
- Belt
- Vines
- Tree branches
- Small bits of food or worms
- Bottle
- Small rocks
- Large rock

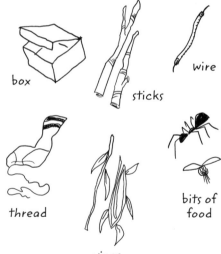

box

sticks

wire

thread

vines

bits of food

What to Do   # BOX TRAP

First, locate a cardboard box or use wood branches tied together with vines or strings to make a sneaky box shape. The box must have a door that can be propped open with a small branch in order to close behind the animal.

Next, position some small food or worms or a shiny item to lure the animal into the box opening. Set the stick so that it is positioned to keep the box door open gently, not rigidly. This allows the animal to bump into it and inadvertently close the door, thereby trapping the animal inside as shown in **Figure 1**.

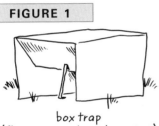

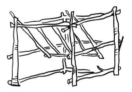

box trap
(flap weighted to close shut)

stick-made box trap
with door

If the box has an open top or no door flap, place the branch so that it props the box up off the ground. Set the bait near the base of the branch so when the curious animal moves about, it will bump into the stick and cause the box to drop on him. See **Figure 2**.

SNARE TRAP

First, locate items like long vines, thread from clothing, string, twine, or wire to make into a sneaky rope. Braid the material for strength and tie knots in both ends.

Next, tie one end of the rope around the base of a tree and secure it with a knot. Make a loop with the other end; this will become the snare. See **Figure 3**.

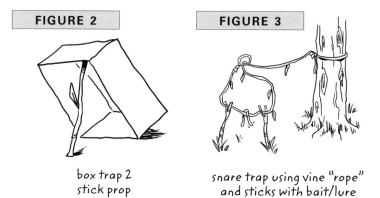

box trap 2
stick prop

snare trap using vine "rope"
and sticks with bait/lure

Hide the loop with grass and twigs placed on top of it. Place small bits of food or worms to attract an animal into the snare's loop. When a small animal runs into the loop, it will attempt to escape and pull it tighter, thus becoming trapped more tightly.

ROCK TRAP

If no box is available, use a large rock to trap a small animal.

As for the box trap, place a branch so it props up the rock. Position the stick very delicately, because it should fall at the lightest touch.

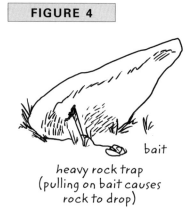

Place bait material next to the base of the branch so that the animal will knock over the stick while investigating and become trapped by the weight of the rock, as shown in **Figure 4**.

heavy rock trap
(pulling on bait causes
rock to drop)

Sneaky Fishing

Experienced outdoorsmen know the value of being able to lure and trap fish. In an emergency survival situation, when you're near water and cannot get back to your campsite, you can use everyday items to make sneaky fishing rigs.

A sneaky fishing rig consists of a line, a hook, and a lure. The line extends the hook and lure deep into the water. The lure attracts the fish, and, with a little luck, the hook ensnares it, allowing you to pull it in for a meal. If the hook and lure stay too close to the water's surface, tie a small rock to the line so it will sink deeper into the water.

What's Needed

- Foil or other shiny object
- Bait (worms, insects, food bits, or feathers)
- Wire
- Strong threads from clothes or string
- Dental floss
- Paper clip
- Vine
- Plastic bottle
- Tree branches

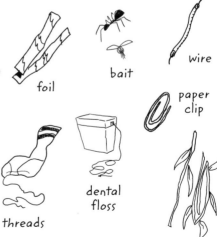

foil

bait

wire

paper clip

threads

dental floss

vines

What to Do
FISHING RIGS

Lines. Fashion your fishing line from items you have on or around you, such as dental floss, clothing thread, or wire. Nature provides tree vines that, when flattened with a rock and braided together, can make a good makeshift fishing line.

Hooks. You can make sneaky fishhooks using pins, needles, wire, small nails—even a thorny branch. Paper clips, a straightened key ring, stiff wire, shells, and bones can also be used.

Lures. Just because a hook is dangling from a line, you can't depend on a fish to investigate it, so some sort of bait is required. Use whatever you can find in your area that can be stuck on the end of the hook. Insects, worms, and small bits of food will do the trick. If you're out of real food, objects like a button, a shiny chain or foil wrapper, a feather, a small key, or even a fish-shaped leaf will increase the odds that you'll lure a fish to swallow the hook.

Simply wrap the line around the hook, place the bait item on the hook, and extend the line as far out and into the water as you can as shown in **Figure 1**. If possible, prop up the line with a forked branch to allow it to extend farther into the water.

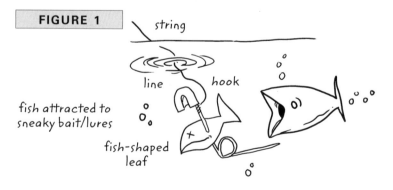

FIGURE 1

string

line   hook

fish attracted to
sneaky bait/lures

fish-shaped
leaf

FISH NETS

If you're near a stream, you may be able to catch fish with a net, which is easy enough to make.

First, find a long tree branch that splits off into a fork and remove all the leaves on the branch. Take a spare shirt and tie the sleeves and collar area into a knot.

Then place the shirt upside down into the forked branch so it produces a makeshift net. Secure the shirt to the branch using whatever you have—paper clips, a key ring, wire, or smaller forked branches—to keep the shirt tight on the branch ends. See **Figure 2**.

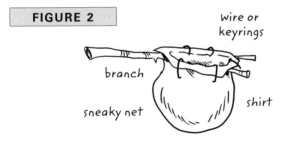

FIGURE 2

wire or keyrings

branch

sneaky net

shirt

Last, lay the branch and shirt into the water, and you can ensnare aquatic creatures in your Sneaky Net.

AQUA TRAPS

A spare plastic bottle can provide another sneaky method to trap small fish if you're near a stream or pond.

First, use a knife or sharp rock to cut off the top of the bottle. Then place some sort of bait material in the bottom of the bottle. You can use insects, worms, small food bits, a button, a shiny chain, a foil wrapper, or a feather.

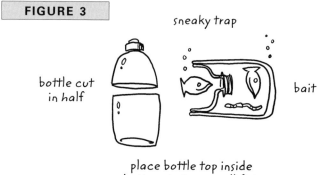

FIGURE 3

sneaky trap

bottle cut
in half

bait

place bottle top inside
bottom to trap small fish

Next, turn the bottle top upside down and push it, mouth down, into the bottom section so it's wedged tight. See **Figure 3**. Now place the bottle near the edge of a stream or pond. Curious fish will swim through the mouth of the bottle looking for the bait but will not be able to get out. Leave as many of these aqua traps as you can in the water, and with a little luck you will find a fish dinner waiting for you on your return.

Sneaky Whistle

If you're trapped in the wilderness and need to signal for assistance, especially during daylight hours, a whistle is far more noticeable then a visual signal. It can mean the difference between being found or left behind.

If you neglected to bring a whistle with you, you can make a sneaky version from things around you easily enough.

What's Needed

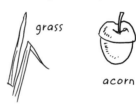

- Long blade of grass
- Hollowed-out acorn
- Small pebble

What to Do

To make a whistle with a long blade of grass, hold it between your thumbs, as shown in **Figure 1**. Blow through the gap between your thumb knuckles. After some practice, and adjustments when necessary, you'll be able to produce a loud whistle with this nature-provided adaptation.

Similarly, you can blow into a variety of small shells, like an acorn, with your cupped hands and achieve a loud piercing sound that will surely be heard by anyone nearby. See **Figure 2**.

If no whistle sound is produced, place a small pebble in the shell and try again. Another method is to cut a tiny hole in the shell with a knife or sharpened rock and then cup your hands to produce a whistle effect.

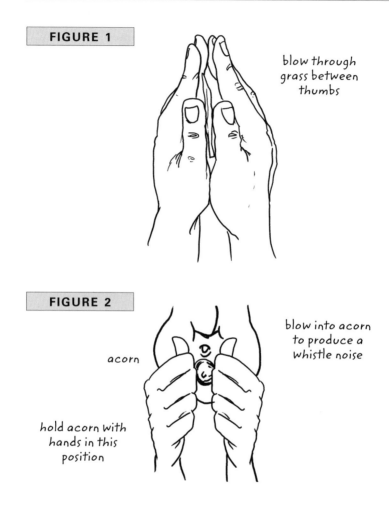

FIGURE 1

blow through
grass between
thumbs

FIGURE 2

acorn

blow into acorn
to produce a
whistle noise

hold acorn with
hands in this
position

Part IV

Gadget Jacket

With so many sneaky project possibilities available here, and in *Sneaky Uses for Everyday Things,* you'll want a way to keep them hidden and handy. The most practical way is to make a portable home for them in your own gadget jacket.

The following projects illustrate how to adapt a favorite jacket to make your sneaky devices instantly available for a variety of purposes. You can mix and match different gadgets according to your "mission," limited only by your imagination and the availability of your devices.

The examples that follow were selected for practicality. For example, typically you can outfit your gadget jacket with a mini remote-control transmitter, a voice recorder, a retractable magnet (to verify currency), and a mini-camera. When traveling, the gadget jacket can be outfitted with a compass, mini-poncho, pocket heater, mini survival kit, telescope, and a defensive repellent sprayer.

Sneaky Interior Pockets

The interior of the gadget jacket has a modular design to allow for both easy expansion and quick removal of the added items (for cleaning, inspection, and while traveling). Don't use the jacket's original pockets for your sneaky devices; they may clink against other personal items. Also, you should know the locations of operational buttons and controls for all the gadgets without having to peer into your exterior pockets. That's only possible when your gadgets are specifically arranged in custom-made pockets.

What's Needed

- Jacket
- Cloth to match jacket lining
- Nylon thread
- Velcro strips

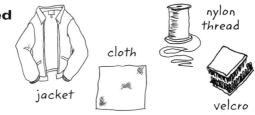

jacket

cloth

nylon thread

velcro strips

What to Do

The areas where interior pockets can be added without unsightly bulges are alongside the upper back, near the underarms, and on the lower back and front. The upper side near the underarm is perfect for longer tube-shaped items, like the safety light stick and the mini air pump.

Custom pockets that provide a perfect fit for easy access and reinsertion of devices can be quickly made with cloth material that matches the jacket lining. Velcro strips that also match the jacket's interior color allow for easy installation and quick removal of the pockets.

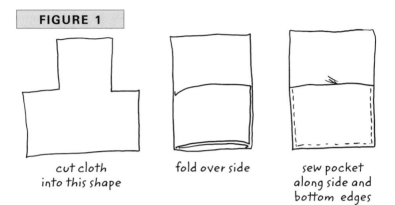

FIGURE 1

cut cloth
into this shape

fold over side

sew pocket
along side and
bottom edges

You should carefully plan which device or group of devices you will place in each particular pocket. Measure the cloth material for a snug fit (but not so snug that it is difficult to insert and remove an item). **Figure 1** shows how to cut a piece of material into a **T** shape and sew it together to create a mini-pocket with flap. Simply sew two Velcro strips to the back of the pocket and to the desired area in the jacket for mounting. See **Figure 2**.

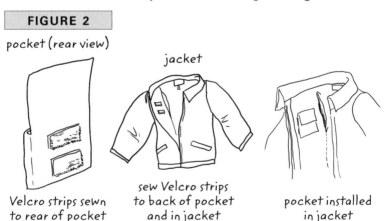

FIGURE 2

pocket (rear view)

jacket

Velcro strips sewn
to rear of pocket

sew Velcro strips
to back of pocket
and in jacket

pocket installed
in jacket

FRONT INTERIOR POCKETS

The lower front interior pockets can store various sneaky items that come in handy in everyday life.

The list includes—but is not limited to—a mini pocket multi-tool, a telescoping antenna with magnet (when the mirror with magnet is attached, it acts as an around-the-corner sneaky viewer), a whistle or siren, mini-telescope, key finder, a mini–universal remote control, and a mini-camera.

Sneaky Camera Cozy. In order to store and use the mini-camera secretly, hide it inside a long candy box, along with a shorter box that's filled with candy.

To build the Sneaky Camera Cozy, position the camera inside a long candy box. Note where the camera shutter is located in relation to the box (if necessary, mark it with a pencil) and cut a hole in the front of the box for the lens to receive light. See **Figure 1**.

Then place a smaller candy box with candy inside in front of the camera. This will fill up the space in the larger box; you can even take candy from it occasionally so as not to raise suspicion. When needed, you can carry the box and position it properly to take a sneaky snapshot. See **Figure 2**.

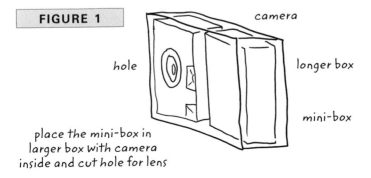

FIGURE 1

camera

hole

longer box

mini-box

place the mini-box in larger box with camera inside and cut hole for lens

press shutter to take
secret, sneaky snapshots

SIDE INTERIOR POCKETS

The lower side interior pockets are perfect for slightly bigger items that need not be frequently accessed, such as the mini-fan and hand warmer.

Other items for side pockets include balloons, a mini air pump (for emergency flotation), and an air breather (for protection in a smoked-filled environment). See **Figure 4**.

BACK INTERIOR POCKETS

The lower back pockets provide storage space for emergency items such as an ultra-thin poncho, an emergency foil blanket, a safety light stick, and various sneaky pens that include survival and security implements. See **Figure 5**.

SHOULDER

To obtain the best long-range distance for your walkie-talkie or radio-control transmitter, sew a thin wire, the same color as the jacket's interior, into the upper shoulder area. See **Figure 6**. Then attach the wire to the antenna connector on the component board of the walkie-talkie or radio-control transmitter.

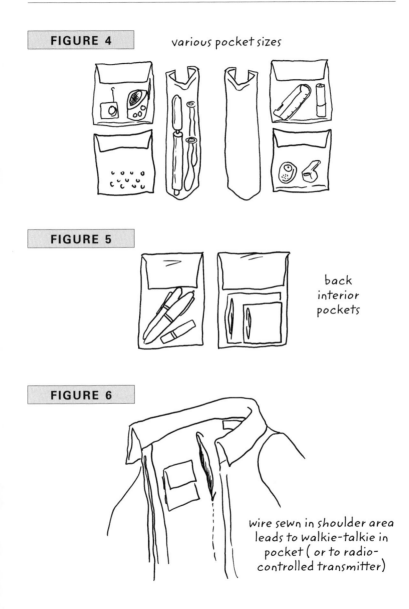

FIGURE 4   various pocket sizes

FIGURE 5

back
interior
pockets

FIGURE 6

wire sewn in shoulder area
leads to walkie-talkie in
pocket (or to radio-
controlled transmitter)

Sneaky Sleeve Pockets

Want instant access to a small defense item (like a repellent sprayer or other device) secretly propelled in the palm of your hand? Once you've made your own sneaky sleeve gadget for just a few everyday items, you'll never want to be without it.

What's Needed

- Jacket
- Cloth
- Nylon thread
- Paper clips
- Woven elastic strip
- Elastic band
- Velcro strips
- Scissors

jacket

cloth

nylon thread

paper clips

scissors

velcro strips

What to Do

By now you should have created and mounted a few pockets in the gadget jacket for your devices. The Sneaky Sleeve Pocket will use a similar design with one difference: The sleeve pocket is mounted upside down with an elastic band and thread for a sneaky release mechanism. If desired, the sneaky pocket can be mounted at the jacket's waist area for storing larger objects.

For this project, the sleeve pocket hides a mini-bottle, as described in the Sneaky Repellent project, but you can substitute another compact device or tool of your choice. Be sure to use cloth material, Velcro strips, an elastic strip, and strong nylon thread that match the jacket's color.

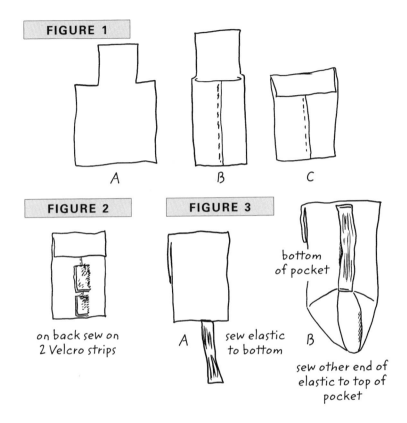

FIGURE 1

A

B

C

FIGURE 2

on back sew on
2 Velcro strips

FIGURE 3

A   sew elastic
to bottom

bottom
of pocket

B

sew other end of
elastic to top of
pocket

First, cut a piece of cloth into a **T** shape and sew it together
to create a mini-pocket with flap that will house the bottle. The
pocket should fit the bottle but not be so tight as to keep it from
easily sliding out of the pocket's bottom hole. Sew two Velcro
strips to the back of the pocket and a matching pair of strips
inside the jacket's sleeve area. See **Figures 1 and 2**.

Then sew a strip of woven elastic (that is half the pocket's
length) to the bottom and top of the pocket, as shown in **Figure 3**.

FIGURE 4

sew nylon thread
to band of elastic

The elastic will cause the bottom of the pocket to curl up, which prevents the bottle from falling out. Next, sew a 6-inch length of nylon thread to the elastic strip at the bottom of the pocket. See **Figure 4**.

Now you can test the quick-release action of the threaded pocket opener. As shown in **Figure 5**, hold the pocket and pull the thread downward. Slip the bottle in the pocket. Let the thread go and the bottom of the pocket curls up, keeping the bottle from falling out. Pull the thread and the bottle should easily slip out. If not, make adjustments to the size of the pocket for a better fit.

FIGURE 5

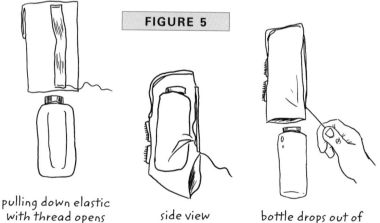

pulling down elastic
with thread opens
bottom

side view
of pocket

bottle drops out of
bottom of pocket

FIGURE 6   **FIGURE 7**

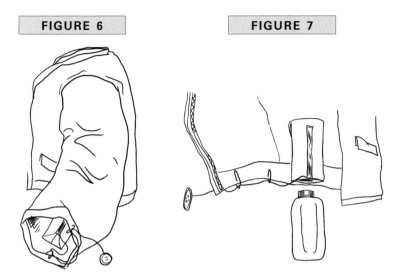

Last, remove the jacket's sleeve button and, using scissors, poke a small hole in the sleeve for the thread to fit through. Place the sneaky pocket on the Velcro strips sewn into the jacket sleeve. Lead the thread through the sleeve hole and cut it so it just reaches the button. Sew the end of the thread to the button. As shown in **Figure 6**, you can sew a paper clip or two in the sleeve as a guide for the thread. This will ensure a proper fit and a smooth release when you pull the button. With the bottle in the pocket and your arm positioned downward, you can pull the sleeve button and the bottle will slide right into the palm of your hand.

Figure 7 illustrates the same sneaky pocket design mounted at the lower side area of the jacket near the waist. Pulling on the jacket's front button causes the thread to open the bottom; your hand should be ready to grab the bottle when it falls.

Sneaky Buttons

Virtually every area of a garment can be used for sneaky subterfuge. Even the gadget jacket's buttons can be used to hide sneaky devices.

What's Needed

- Magnet
- Retractable key fob
- Buttons
- Button caps
- Velcro strips (optional)
- Paper currency

magnets

buttons

Velcro strips

What to Do

You can use the jacket's existing buttons or sew on a different type, depending on your needs. Or you can sew cloth over an item, like a compass or magnet, and sew it on an existing button as shown in **Figure 1**.

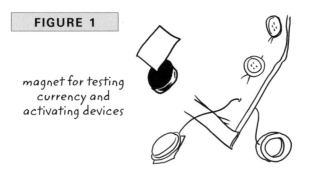

FIGURE 1

magnet for testing
currency and
activating devices

FIGURE 2

retractable
key fob

For an added sneaky effect, use Velcro strips to attach a mini-retractable key fob to the inside of the jacket and connect its cord to a button on the jacket's exterior, as shown in **Figure 2**.

Whenever you doubt the authenticity of paper currency, simply pull your button magnet and perform the magnetic attraction test: Fold the bill in half and set it upright on a table, as in **Figure 3**. Point the button magnet near the edge of the bill but do not touch it. A legitimate bill will move toward the magnet. See **Figure 4**.

FIGURE 3

fold bill and
set upright
on table

FIGURE 4

real bill
moves toward
button magnet

button
magnet

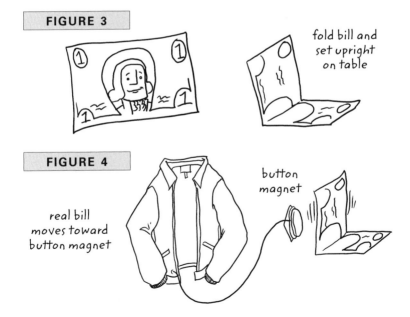

Sneaky Listener

Even when you're not wearing the gadget jacket, it can provide you with valuable information. A pair of mini walkie-talkies can be used to provide your gadget jacket with a sneaky communications system.

What's Needed

- Jacket
- Pair of mini walkie-talkies
- Candy box
- Tape (plastic or duct)

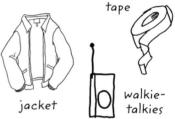

What to Do

First, apply tape to cover the walkie-talkie's TALK button so it is always in the TRANSMIT mode. Then position the walkie-talkie in a candy box, punch a small hole at the location of the ON/OFF button, and put the candy box in your jacket pocket. In this manner, you can reach in and turn on the walkie-talkie while it's in the candy box. (**Figures 1 and 2**).

As shown in **Figure 3**, once you activate the ON/OFF switch and leave the jacket in a suitable location, you can use the other walkie-talkie (stored on your belt clip or in your pocket) to monitor conversations in the jacket's vicinity.

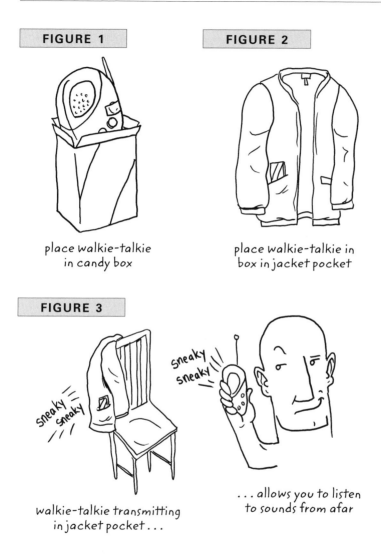

FIGURE 1

place walkie-talkie
in candy box

FIGURE 2

place walkie-talkie in
box in jacket pocket

FIGURE 3

sneaky
sneaky

sneaky
sneaky

walkie-talkie transmitting
in jacket pocket...

...allows you to listen
to sounds from afar

Sneaky Collar

The gadget jacket's collar provides a mounting location for a couple of useful sneaky devices.

What's Needed

- Jacket
- Velcro strips
- Flexible mini-light
- Two small magnets
- Small compact mirror
- Mini voice recorder

jacket

Velcro strips

magnets

What to Do

First, sew two small Velcro squares on each side of the jacket under the collar area. This will allow you to quickly position and mount a variety of miniature devices that are available for your eyes and ears. Simply apply the other half of the Velcro squares, with its double-stick tape material, to the items of your choice. See **Figure 1**.

For example, as shown in **Figure 2**, you can mount a compact light with a flexible end to have hands-free access to a light source when required.

Want to see behind you while walking down the street or at an ATM? Glue a small magnet on the side of the flexible light's end and to a small compact mirror. Mount the mirror on the end of the shaft. This will provide you with an adjustable mount for a sneaky rearview mirror. See **Figure 2**.

Figure 3 illustrates how a mini voice recorder can be mounted under the collar, allowing you the ability to secretly record important conversations. Or you can mount the voice recorder in another location in the jacket, such as the front pocket or sleeve.

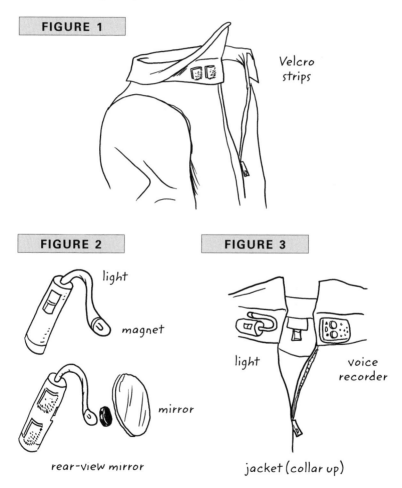

FIGURE 1

Velcro strips

FIGURE 2

light

magnet

mirror

rear-view mirror

FIGURE 3

light

voice recorder

jacket (collar up)

Portable Heater

If you've ever wished for a portable heater in a cold-weather situation, wish no longer. Using everyday items, you can make a compact portable heater that will distribute warm air to every part of your gadget jacket (and optionally your pants too).

What's Needed

- Hand-warmer heat pack (sold at drug and sporting goods stores)
- Motor from toy car or battery-powered mini-fan
- Wire
- Cardboard
- Tape

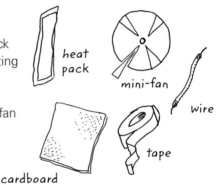

heat pack

mini-fan

wire

cardboard

tape

What to Do

Remove a motor from a toy car and tape a cardboard-made fan to its shaft. See **Figure 1**. While the fan is in the jacket, you'll need to protect its blades so they can spin freely without touching other items.

To make a fan blade protector, cut out a triangular cone shape from the cardboard and tape it to the body of the fan as shown in **Figure 2**. Turn on the fan and, if necessary, position it so the blades spin freely without touching the cardboard.

Then cut small holes in an interior pocket of the jacket (or sew another pocket inside the lining) for ventilation and place the hand warmer and fan inside. When you're cold, activate the hand

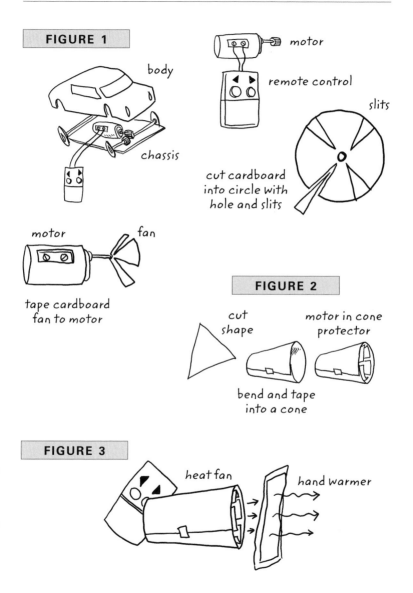

FIGURE 1

body

chassis

motor

remote control

slits

cut cardboard into circle with hole and slits

motor   fan

tape cardboard fan to motor

FIGURE 2

cut shape

motor in cone protector

bend and tape into a cone

FIGURE 3

heat fan

hand warmer

FIGURE 4

hand warmer

motor/fan

optional: straw
led down to
pants

pocket with
vent holes

warmer (follow the manufacturer's instructions) and position it
near the ventilation holes so the fan will blow on it and the heat
will travel from the holes to the jacket's interior, as shown in
Figure 3.

Additionally, a long flexible straw or tube can be attached to
the fan enclosure and led down the back of your pants to warm
up your lower extremities. See **Figure 4**.

GOING FURTHER

The gadget jacket projects will undoubtedly provide you with
sneaky fun ideas for your own personal design. **Figure 5** illustrates
how all of the items will appear when installed in the jacket, along
with other devices that complement the jacket's devices (like an
additional walkie-talkie, sneaky pen, power ring, defense ring,
sneaky watch, and sneaky shoestrings) from projects in this
book and the earlier *Sneaky Uses for Everyday Things*.

Have fun devising and making your own sneaky jacket adaptations and be sure to check for additional ideas (and post your
own) at www.Sneakyuses.com

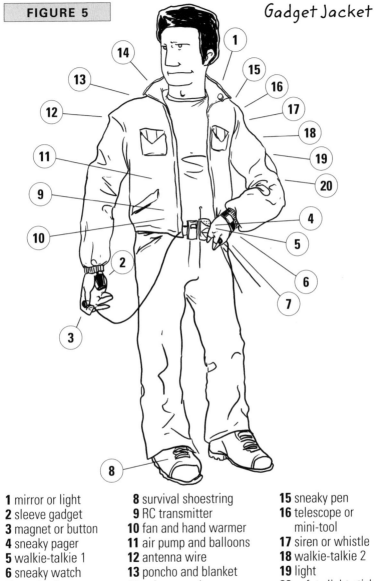

FIGURE 5

Gadget Jacket

1 mirror or light
2 sleeve gadget
3 magnet or button
4 sneaky pager
5 walkie-talkie 1
6 sneaky watch
7 sneaky ring

8 survival shoestring
9 RC transmitter
10 fan and hand warmer
11 air pump and balloons
12 antenna wire
13 poncho and blanket
14 voice recorder

15 sneaky pen
16 telescope or mini-tool
17 siren or whistle
18 walkie-talkie 2
19 light
20 safety light stick

Resources

WEB SITES

Science Sites
amasci.com/
build-it-yourself.com
discovercircuits.com
exploratorium.edu
howtoons.net/
kidsinvent.org
makezine.com
sciencetoymaker.org
sneakyuses.com
theteachersguide.com/QuickScienceActivities.html
us.brainium.com
uspto.gov/go/kids
wildplanet.com

Frugal and Thrift Sites
choose2reuse.org
freegiftclub.net
frugalcorner.com
frugalitynetwork.com
getfrugal.com
make-stuff.com
ready-made.com
Recycle.net
thefrugalshopper.com
thriftydeluxe.com
wackyuses.com
watchthepennies.com

Gadget Sites

advanced-intelligence.com

berberblades.com

casio.com

colibri.com

dailygadget.com

equalizers1.com

girltech.com

gizmodo.com

ijustgottahavethat.com

inventorsdigest.com

johnson-Smith.com

leatherman.com

netgadget.net

nutsandvolts.com

popgadget.net

recorderracer.org

robotstore.com

scientificsonline.com

smartplanet.net

spyderco.com

spy-gear.net

spymall.com

swissarmy.com

swisstechtools.com

the-gadgeteer.com/cgi-bin/redirect.cgi/gadget

thinkgeek.com

topeak.com

Survival Sites

americansurvivalist.com
backwoodshome.com
backwoodsmanmag.com
basegear.com
beprepared.com
campmor.com
emergencypreparednessgear.com
equipped.com
fieldandstream.com
hikercentral.com/survival
homepower.com
productsforanywhere.com
ruhooked.com
secretsofsurvival.com
self-reliance.net
simply-survival.com
skillsofsurvival.com
survival-center.com
Survival.com
Survivaliq.com
survivalplus.com
Survivalx.com
wildernesssurvival.com
wilderness-survival.net
windpower.org

Home Security Sites

mcgruff.org
ncpc.org
safesolutionsystems.com
X10.com
youdoitsecurity.com

Science and Technology Sites

about.com

boydhouse.com/crystalradio

Craftsitedirectory.com

discover.com

hallscience.com

HomeAutomationMag.com

howstuffworks.com

Johnson-Smith.com

midnightscience.com

RadioShack.com

sciencekits.com

scienceproject.com

Scientificsonline.com

scitoys.com

thinkgeek.com

wildplanet.com

Other Web Sites of Interest

Almanac.com

Doityourself.com

Movie-Mistakes.com

Nitpickers.com

Oopsmovies.com

Popsci.com

Popularmechanics.com

rube-goldberg.com

Smarthome.com

tbotech.comrotorsportz.com

Thefunplace.com

Tipking.com

Toollogic.com

Recommended Reading

Books

David Borgenicht & Joe Borgenicht, *The Action Hero's Handbook* (Quirk Books)

Robert Young Decton, *Come Back Alive* (Doubleday)

Department of the Air Force, *US Air Force Search & Rescue Handbook* (Lyons Press)

Department of the Army, *US Army Survival Handbook* (Lyons Press)

Simon Field, *Gonzo Gizmos* (Chicago Review Press)

Ira Flatow, *They All Laughed . . . From Light Bulbs to Lasers: The Fascinating Stories Behind the Great Inventions That Have Changed Our Lives* (Harper Perennial)

Joey Green, *Clean It! Fix It! Eat It!: Easy Ways to Solve Everyday Problems with Brand-Name Products You've Already Got Around the House* (Prentice Hall)

―――, *Clean Your Clothes with Cheez Wiz: And Hundreds of Offbeat Uses for Dozens More Brand-Name Products* (Prentice Hall)

―――, *Joey Green's Encyclopedia of Offbeat Uses for Brand-Name Products* (Prentice Hall)

―――, *The Mad Scientist Handbook II* (Perigee Trade)

Lois H. Gresh & Robert Weinberg, *The Science of Superheroes* (John Wiley & Sons)

William Gurstelle, *Backyard Ballistics* (Chicago Review Press)

Garth Hattingh, *The Outdoor Survival Handbook* (New Holland Publishers)

Dave Hrynkiw and Mark W. Tilden, JunkBots, Bugbots, and Bots on Wheels: Building Simple Robots With BEAM Technology (McGraw-Hill Osborne Media)

Vicky Lansky, *Another Use For 101 Common Household Items* (Book Peddlers)

————, *Baking Soda: Over 500 Fabulous, Fun, and Frugal Uses* (Book Peddlers)

————, *Don't Throw That Out: A Pennywise Parent's Guide* (Book Peddlers)

————, *Transparent Tape: Over 350 Super, Simple, and Surprising Uses* (Book Peddlers)

Joel Levy, *Really Useful: The Origins of Everyday Things* (Firefly Books)

Hugh McManners, *The Complete Wilderness Training Book* (Dorling Kindersley)

Forrest M. Mims III, *Circuits and Projects* (Radio Shack)

————, *Science and Communications Circuits and Projects* (Radio Shack)

Steven W. Moje, *Paper Clip Science* (Sterling Publishing Co.)

Bob Newman, *Wilderness Wayfinding: How to Survive in the Wilderness as You Travel* (Paladin Press)

Tim Nyberg and Jim Berg, *The Duct Tape Book* (Workman Publishing Co., 1994)

————, *Duct Tape Book Two: Real Stories* (Pfeifer-Hamilton Publishing, 1995)

————, *The Jumbo Duct Tape Book* (Workman Publishing Co., 2000)

Larry Dean Olsen, *Outdoor Survival Skills* (Chicago Review Press)

Joshua Piven & David Borgenicht, *The Worst Case Scenario Survival* (Chronicle Books)

————, *The Worst Case Scenario Travel* (Chronicle Books)

Royston M. Roberts, *Serendipity* (John Wiley & Sons)

Jim Wilkinson and Neil A. Downie, *Vacuum Bazookas, Electric Rainbow Jelly, and 27 Other Saturday Science Projects* (Princeton University Press)

John Wiseman, *The SAS Survival Handbook* (Harvill Books)

Magazines

Ready Made
Backpacker
Outside
Outdoor Life
Mother Earth News
E Magazine
Self-Defense
Popular Science
Popular Mechanics
Poptronics
Nuts and Volts
Make